INTRODUCING

Quantum Theory

J.P. McEvoy • Oscar Zarate

Edited by Richard Appignanesi

Icon Books UK   Totem Books USA

This edition published in the UK in 2004 by Icon Books Ltd., The Old Dairy, Brook Road, Thriplow, Cambridge SG8 7RG email: info@iconbooks.co.uk www.iconbooks.co.uk

This edition published in the USA in 2004 by Totem Books Inquiries to: Icon Books Ltd., The Old Dairy, Brook Road, Thriplow, Cambridge SG8 7RG, UK

Sold in the UK, Europe, South Africa and Asia by Faber and Faber Ltd., 3 Queen Square, London WC1N 3AU or their agents

Distributed to the trade in the USA by National Book Network Inc., 4720 Boston Way, Lanham, Maryland 20706

Distributed in the UK, Europe, South Africa and Asia by TBS Ltd., Frating Distribution Centre, Colchester Road, Frating Green, Colchester CO7 7DW

Distributed in Canada by Penguin Books Canada, 90 Eglinton Avenue East, Suite 700, Toronto, Ontario M4P 2YE

This edition published in Australia in 2004 by Allen and Unwin Pty. Ltd., PO Box 8500, 83 Alexander Street, Crows Nest, NSW 2065

ISBN 1 84046 577 8

Previously published in the UK and Australia in 1996 under the title *Quantum Theory for Beginners* and in 1999 under the current title

Reprinted 1997, 1998 (twice), 1999, 2000, 2001, 2002, 2003, 2005

Originating editor: Richard Appignanesi

Printed and bound in Malta
by Gutenberg Press Ltd

What is Quantum Theory?

Quantum theory is the most successful set of ideas ever devised by human beings. It explains the periodic chart of the elements and why chemical reactions take place. It gives accurate predictions about the operation of lasers and microchips, the stability of DNA and how alpha particles tunnel out of the nucleus.

Niels Bohr's presentation of quantum theory in 1927 remains today's orthodoxy. But Einstein's *thought* experiments in the 1930s questioned the theory's fundamental validity and are still debated today. Could he be right again? Is there something missing?

Let's begin at the beginning . . .

Introducing Quantum Theory . . .

The problem is this. Just before the turn of the century, physicists were so absolutely certain of their ideas about the nature of matter and radiation that any new concept which contradicted their **classical** picture would be given little consideration.

Not only was the mathematical formalism of **Isaac Newton** (1642–1727) and **James Clerk Maxwell** (1831–79) impeccable, but predictions based on their theories had been confirmed by careful detailed experiments for many years. The Age of Reason had become the age of certainty!

Classical Physicists

What is the definition of "classical"?

By **classical** is meant those late 19th century physicists nourished on an academic diet of Newton's mechanics and Maxwell's electromagnetism – the two most successful syntheses of physical phenomena in the history of thought.

WITH A SIMPLE INCLINED PLANE AND A METAL SPHERE, I DEMONSTRATED THAT THE GREAT **ARISTOTLE'S** PHYSICS WAS FLAWED.

OH, STOP SHOWING OFF!

Testing theories by observation had been the hallmark of good physics since **Galileo** (1564–1642). He showed how to devise experiments, make measurements and compare the results with the predictions of mathematical laws.

The interplay of theory and experiment is still the best way to proceed in the world of acceptable science.

It's All Proven (and Classical) . . .

During the 18th and 19th centuries, Newton's laws of motion had been scrutinized and confirmed by reliable tests.

MY GRAVITATION LAW HAS BEEN USED TO PREDICT MEASURED MOVEMENTS OF THE PLANETS WITH GREAT ACCURACY . . .

I PREDICTED THE EXISTENCE OF INVISIBLE "LIGHT" WAVES IN MY ELECTROMAGNETIC WAVE THEORY OF 1865, AND *HEINRICH HERTZ* (1857-94) DETECTED THE SIGNALS IN 1888 IN HIS BERLIN LABORATORY. NOW THEY'RE CALLED *RADIO WAVES*.

THESE WAVES REFLECT AND REFRACT JUST LIKE LIGHT. MAXWELL WAS RIGHT.

No wonder these classical physicists were confident in what they knew!

"Fill in the Sixth Decimal Place"

A classical physicist from Glasgow University, the influential **Lord Kelvin** (1824–1907), spoke of only two dark clouds on the Newtonian horizon.

HOW WAS I TO KNOW THAT ONE OF THESE CLOUDS WOULD DISAPPEAR ONLY WITH THE ADVENT OF RELATIVITY — AND THE OTHER WOULD LEAD TO QUANTUM THEORY?

In June 1894, the American Nobel Laureate, **Albert Michelson** (1852–1931), thought he was paraphrasing Kelvin in a remark which he regretted for the rest of his life.

ALL THAT REMAINS TO DO IN PHYSICS IS FILL IN THE SIXTH DECIMAL PLACE. (I CAN'T BELIEVE I SAID THAT!)

0.12345...

The Fundamental Assumptions of Classical Physics

Classical physicists had built up a whole series of assumptions which focused their thinking and made the acceptance of new ideas very difficult. Here's a list of **what they were sure of** about the material world . . .

1) The universe was like a giant machine set in a framework of absolute time and space. Complicated movement could be understood as a simple movement of the machine's inner parts, even if these parts can't be visualized.

*2) The Newtonian synthesis implied that all motion had a **cause**. If a body exhibited motion, one could always figure out what was producing the motion. This is simply **cause and effect**, which nobody really questioned.*

*3) If the state of motion was known at one point – say the present – it could be determined at any other point in the future or even the past. Nothing was uncertain, only a consequence of some earlier cause. This was **determinism**.*

*4) The properties of light are **completely described** by Maxwell's electromagnetic **wave** theory and confirmed by the interference patterns observed in a simple double-slit experiment by Thomas Young in 1802.*

*5) There are two physical models to represent energy in motion: one a **particle**, represented by an impenetrable sphere like a billiard ball, and the other a **wave**, like that which rides towards the shore on the surface of the ocean. They are mutually exclusive, i.e. energy must be either one or the other.*

6) It was possible to measure to any degree of accuracy the properties of a system, like its temperature or speed. Simply reduce the intensity of the observer's probing or correct for it with a theoretical adjustment. Atomic systems were thought to be no exception.

Classical physicists believed all these statements to be **absolutely true**. But **all** six assumptions would eventually prove to be **in doubt**. The first to know this were the group of physicists who met at the Metropole Hotel in Brussels on 24 October 1927.

The Solvay Conference 1927 – Formulation of Quantum Theory

A few years before the outbreak of World War I, the Belgian industrialist **Ernest Solvay** (1838–1922) sponsored the first of a series of international physics meetings in Brussels. Attendance at these meetings was by special invitation, and participants – usually limited to about 30 – were asked to concentrate on a pre-arranged topic.

The first five meetings held between 1911 and 1927 chronicled in a most remarkable way the development of 20th century physics. The 1927 gathering was devoted to quantum theory and attended by no less than **nine** theoretical physicists who had made fundamental contributions to the theory. Each of the nine would eventually be awarded a Nobel Prize for his contribution.

This photograph of the 1927 Solvay Conference is a good starting point for introducing the principal players in the development of the most modern of all physical theories. Future generations will marvel at the compressed time scale and geographical proximity which brought these giants of quantum physics together in 1927.

There is hardly any period in the history of science in which so much has been clarified by so few in so short a time.

Look at the sad-eyed **Max Planck** (1858–1947) in the front row next to **Marie Curie** (1867–1934). With his hat and cigar, Planck appears drained of vitality, exhausted after years of trying to refute his own revolutionary ideas about matter and radiation.

A few years later in 1905, a young patent clerk in Switzerland named **Albert Einstein** (1879–1955) generalized Planck's notion.

That's Einstein in the front row centre, sitting stiffly in his formal attire. He had been brooding for over twenty years about the quantum problem without any real insights since his early 1905 paper. All the while, he continued to contribute to the theory's development and endorsed original ideas of others with uncanny confidence. His greatest work – the General Theory of Relativity – which had made him an international celebrity, was already a decade behind him.

In Brussels, Einstein had debated the bizarre conclusions of the quantum theory with its most respected and determined proponent, the "great Dane" **Niels Bohr** (1885–1962). Bohr – more than anyone else – would become associated with the struggle to interpret and understand the theory. At the far right of the photo, in the middle row, he is relaxed and confident – the 42 year old professor at the peak of his powers.

IN MY LECTURE, I REVIEWED THE PROBABILISTIC INTERPRETATION OF QUANTUM THEORY TO THE APPARENT SATISFACTION OF MOSTLY EVERYONE, EXCEPT **EINSTEIN.**

Thus began a running argument between these two masters of 20th century physics, lasting up to Einstein's death in 1955.

In the back row behind Einstein, **Erwin Schrödinger** (1887–1961) looks conspicuously casual in his sports jacket and bow tie. To his left but one are the "young Turks", **Wolfgang Pauli** (1900–58) and **Werner Heisenberg** (1901–76) – still in their twenties – and in front of them, **Paul Dirac** (1902–84), **Louis de Broglie** (1892–1987), **Max Born** (1882–1970) and Bohr. These men are today immortalized by their association with the fundamental properties of the microscopic world: the *Schrödinger wave equation;* the *Pauli exclusion principle;* the *Heisenberg uncertainty relation,* the *Bohr Atom* . . . and so forth.

They were all there – from Planck, the oldest at 69 years, who started it all in 1900 – to Dirac, the youngest at 25 years, who completed the theory in 1928.

The day after this photograph was taken – 30 October 1927 – with the historic exchanges between Bohr and Einstein still buzzing in their minds, the conferees boarded trains at the Brussels Central Station to return to Berlin, Paris, Cambridge, Göttingen, Copenhagen, Vienna and Zürich.

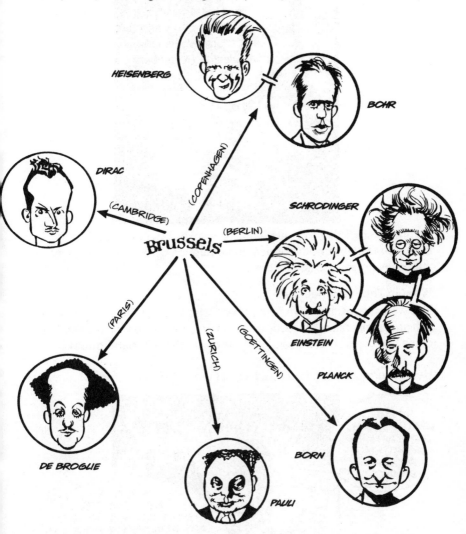

They were taking with them the most bizarre set of ideas ever concocted by scientists. Secretly, most of them probably agreed with Einstein that this madness called the quantum theory was just a step along the way to a more complete theory and would be overthrown for something better, something more consistent with common sense.

But how did the quantum theory come about? What experiments compelled these most careful of men to ignore the tenets of classical physics and propose ideas about nature that violated common sense?

Before we study these experimental paradoxes, we need some background in **thermodynamics** and **statistics** which are fundamental to the development of quantum theory.

What is Thermodynamics?

The word means the *movement of heat,* which always flows from a body of higher temperature to a body of lower temperature, until the temperatures of the two bodies are the same. This is called **thermal equilibrium**.

Heat is correctly described as a **form of vibration** . . .

The First Law of Thermodynamics

Mechanical models to explain the flow of heat developed quickly in 19th century Britain, building on the achievements of **James Watt** (1736–1819), a Scot who had built a working steam engine.

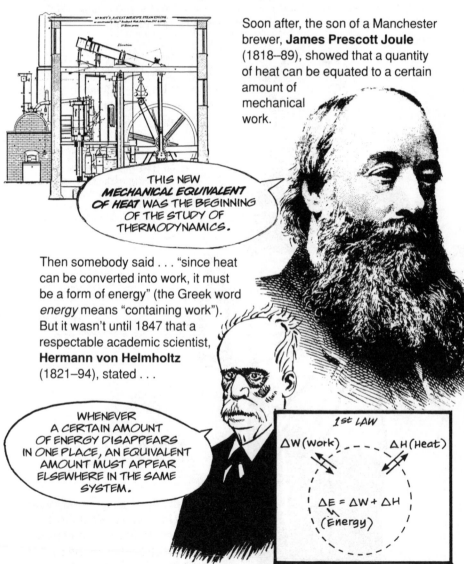

Soon after, the son of a Manchester brewer, **James Prescott Joule** (1818–89), showed that a quantity of heat can be equated to a certain amount of mechanical work.

THIS NEW *MECHANICAL EQUIVALENT OF HEAT* WAS THE BEGINNING OF THE STUDY OF THERMODYNAMICS.

Then somebody said . . . "since heat can be converted into work, it must be a form of energy" (the Greek word *energy* means "containing work"). But it wasn't until 1847 that a respectable academic scientist, **Hermann von Helmholtz** (1821–94), stated . . .

WHENEVER A CERTAIN AMOUNT OF ENERGY DISAPPEARS IN ONE PLACE, AN EQUIVALENT AMOUNT MUST APPEAR ELSEWHERE IN THE SAME SYSTEM.

1st LAW

ΔW (Work) --- ΔH (Heat)

$\Delta E = \Delta W + \Delta H$ (Energy)

This is called the **law of the conservation of energy**. It remains a foundation of modern physics, unaffected by modern theories.

Rudolf Clausius: Two Laws

In 1850, the German physicist **Rudolf Clausius** (1822–88) published a paper in which he called the energy conservation law *The First Law of Thermodynamics.* At the same time, he argued that there was a **second** principle of thermodynamics in which there is always some *degradation* of the total energy in the system, some non-useful heat in a thermodynamic process.

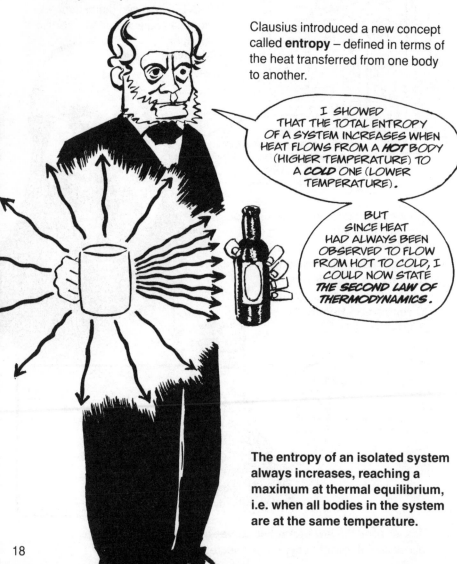

Clausius introduced a new concept called **entropy** – defined in terms of the heat transferred from one body to another.

I SHOWED THAT THE TOTAL ENTROPY OF A SYSTEM INCREASES WHEN HEAT FLOWS FROM A *HOT* BODY (HIGHER TEMPERATURE) TO A *COLD* ONE (LOWER TEMPERATURE).

BUT SINCE HEAT HAD ALWAYS BEEN OBSERVED TO FLOW FROM HOT TO COLD, I COULD NOW STATE *THE SECOND LAW OF THERMODYNAMICS*.

The entropy of an isolated system always increases, reaching a maximum at thermal equilibrium, i.e. when all bodies in the system are at the same temperature.

The Existence of Atoms

A Greek philosopher named **Democritus** (c. 460–370 B.C.) first proposed the concept of atoms (means "indivisible" in Greek).

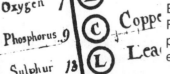

ATOMS ARE THE BASIC BUILDING BLOCKS OF MATTER.

The idea was questioned by Aristotle and debated for hundreds of years before the English chemist **John Dalton** (1766–1844) used the atomic concept to predict the chemical properties of elements and compounds in 1806.

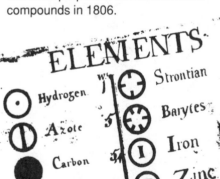

ELEMENTS

Hydrogen

Azote   5

Carbon   5½

Oxygen   7

Phosphorus 9

Sulphur   13

Magnesia   20

Lime   24

Soda   28

Strontian

Barytes

Iron

Zinc

Coppe

Lea

Silv

Go

Plat

Mercury   167

But it was not until a century later that a theoretical calculation by Einstein and experiments by the Frenchman **Jean Perrin** (1870–1942) persuaded the sceptics to accept the existence of atoms as a fact.

However, during the 19th century, even without physical proof of atoms, many theorists used the concept.

19

Averaging Diatomic Molecules

The Scottish physicist J.C. Maxwell, a confirmed atomist, developed his kinetic theory of gases in 1859.

I PICTURED THE GAS TO CONSIST OF BILLIONS OF MOLECULES MOVING RAPIDLY AT RANDOM, COLLIDING WITH EACH OTHER AND WITH THE WALLS OF THE CONTAINER.

SAND
Sand crystals formed from billions of atoms

WATER H_2O
Water molecule formed from three atoms

GAS H_2, O_2 or N_2
Gas molecule formed from two atoms

DNA
DNA molecule formed from hundreds of atoms

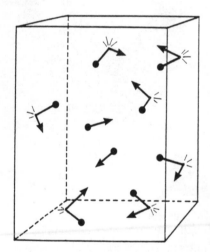

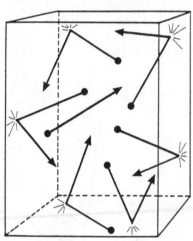

This was qualitatively consistent with the physical properties of gases, if we accept the notion that heating causes the molecules to move faster and bang into the container walls more frequently.

Maxwell's theory was based on *statistical averages* to see if the macroscopic properties (that is, those properties that can be measured in a laboratory) could be predicted from a microscopic model for a collection of gas molecules.

20

Maxwell made four assumptions:

This last assumption was the most unusual and revolutionary, showing a great deal of physical insight by Maxwell.

(600,000,000,000,000,000,000,000).

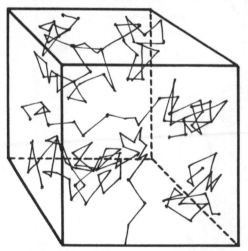

Random Motion seen by Perrin

It would be impossible to try to compute the individual motions of so many particles. But Maxwell's analysis, based on Newton's mechanics, showed that temperature is a measure of the microscopic **mean squared velocity** of the molecules. That is, the average velocity multiplied by itself.

22   *Heat is thus caused by the ceaseless random motion of atoms.*

The real importance of Maxwell's theory is the prediction of the *probable* velocity distribution of the molecules, based on his model. In other words, this gives the *range* of velocities . . . how the whole collection deviates from the average.

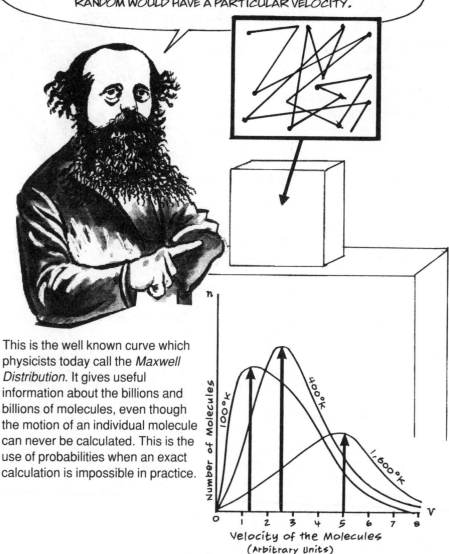

BY ASSUMING THAT THE GAS PARTICLES WERE MOVING UNIFORMLY IN SPACE, WERE MUTUALLY INDEPENDENT AND HAD NO PREFERRED DIRECTION, I COULD COMPUTE THE PROBABILITY THAT A MOLECULE CHOSEN AT RANDOM WOULD HAVE A PARTICULAR VELOCITY.

This is the well known curve which physicists today call the *Maxwell Distribution*. It gives useful information about the billions and billions of molecules, even though the motion of an individual molecule can never be calculated. This is the use of probabilities when an exact calculation is impossible in practice.

Number of Molecules

100°K

400°K

1,600°K

Velocity of the Molecules
(Arbitrary Units)

Ludwig Boltzmann and Statistical Mechanics

In the 1870s, **Ludwig Boltzmann** (1844–1906) – inspired by Maxwell's kinetic theory – made a theoretical pronouncement.

● He presented a general probability distribution law called the **canonical or orthodox distribution**, which could be applied to any collection of entities which have freedom of movement, are independent of each other and interact randomly.

● He formalized the **theorem of the equipartition of energy**.

This means that the energy will be shared equally among all degrees of freedom if the system reaches thermal equilibrium.

● He gave a new interpretation of the Second Law.

When energy in a system is degraded (as Clausius said in 1850), the atoms in the system become more disordered and the entropy increases. But a measure of the disorder can be made. It is the probability of the particular system – defined as the number of ways it can be assembled from its collection of atoms.

ICE (solid)

WATER (steam)

STEAM (Gas)

More precisely, the entropy is given by:

$$S = k \, Log \, W \ldots$$

where **k** is a constant (now called *Boltzmann's constant*) and **W** is the probability that a particular arrangement of atoms will occur.

This work made Boltzmann the creator of *statistical mechanics*, a method in which the properties of macroscopic bodies are predicted by the statistical behaviour of their constituent microscopic parts.

Thermal Equilibrium and Fluctuations

I ASSUMED THAT A SYSTEM WILL EVOLVE FROM A LESS PROBABLE STATE TO A MORE PROBABLE STATE WHEN AGITATED BY HEAT OR MECHANICAL VIBRATION, UNTIL THERMAL EQUILIBRIUM IS REACHED. AT EQUILIBRIUM, THE SYSTEM WILL BE IN ITS MOST PROBABLE STATE WHEN THE ENTROPY IS A MAXIMUM.

IT'S IMPOSSIBLE TO CALCULATE THE MOTION OF BILLIONS AND BILLIONS OF PARTICLES. BUT THE PROBABILITY METHOD CAN GIVE DIRECT ANSWERS FOR THE MOST PROBABLE STATE.

I ALSO INTRODUCED THE CONTROVERSIAL NOTION OF *FLUCTUATIONS*.

A SMALL PROBABILITY EXISTS THAT ALL THE MOLECULES OF A SYSTEM OF CONFINED GAS MIGHT APPEAR FOR AN INSTANT IN JUST ONE CORNER OF THE CONTAINER. THIS POSSIBILITY MUST EXIST IF THE PROBABILISTIC INTERPRETATION OF THE ENTROPY IS TO BE ALLOWED. THIS IS CALLED AN ENERGY FLUCTUATION.

These new ideas – using probabilities and statistics of *microscopic* systems to predict the *macroscopic* properties which can be measured in the laboratory (like temperature, pressure, etc.) – underlie all of what was to come in quantum theory.

The Thirty Years War (1900–30) – Quantum Physics Versus Classical Physics

Now let's look at three critical experiments in the pre-quantum era which could not be explained by a straightforward application of classical physics.

BLACK-BODY RADIATION AND THE ULTRAVIOLET CATASTROPHE (PLANCK'S QUANTUM)

THE PHOTOELECTRIC EFFECT (EINSTEIN'S LIGHT PARTICLES)

BRIGHT LINE OPTICAL SPECTRA (BOHR'S ATOM)

Each involved the interaction of radiation and matter as reported by reliable, experimental scientists. The measurements were accurate and reproducible, yet paradoxical . . . the kind of situation a good theoretical physicist would die for.

We will describe each experiment step-by-step, pointing out the crisis engendered and the solution advanced by Max Planck, Albert Einstein and Niels Bohr respectively. In putting forward their solutions, these scientists made the first fundamental contributions to a new understanding of nature. Today the combined work of these three men, culminating in the Bohr model of the atom in 1913, is known as the *Old Quantum Theory*.

Black-Body Radiation

When an object is heated, it emits radiation consisting of electromagnetic waves, i.e. *light*, with a broad range of frequencies.

MEASUREMENTS MADE ON THE RADIATION ESCAPING FROM A SMALL HOLE IN A CLOSED HEATED OVEN — WHICH IN **GERMANY** WE CALL A ***CAVITY*** — SHOWS THAT THE INTENSITY OF THE RADIATION VARIES VERY STRONGLY WITH THE FREQUENCY OF THE RADIATION.

The dominant frequency shifts to a higher value as the temperature is increased, as shown in the graph drawn from measurements made in the late 19th century.

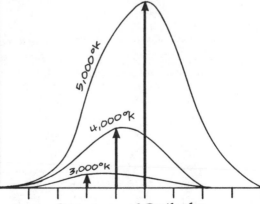

A "Box" (Cavity) of Radiation

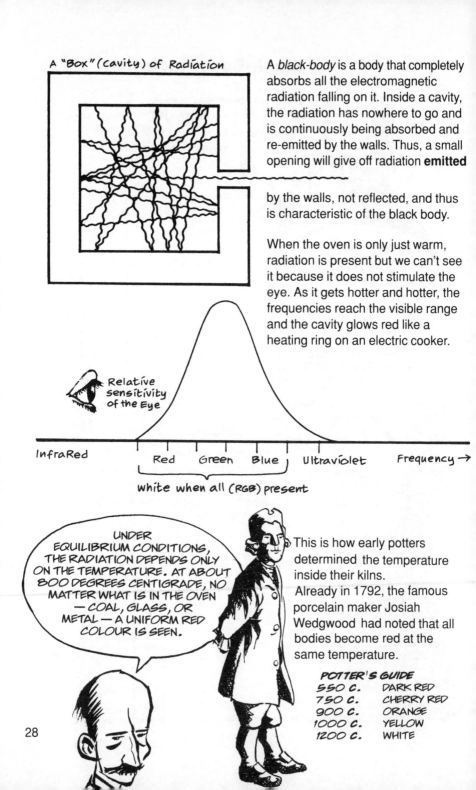

A *black-body* is a body that completely absorbs all the electromagnetic radiation falling on it. Inside a cavity, the radiation has nowhere to go and is continuously being absorbed and re-emitted by the walls. Thus, a small opening will give off radiation **emitted** by the walls, not reflected, and thus is characteristic of the black body.

When the oven is only just warm, radiation is present but we can't see it because it does not stimulate the eye. As it gets hotter and hotter, the frequencies reach the visible range and the cavity glows red like a heating ring on an electric cooker.

Relative sensitivity of the Eye

InfraRed — Red — Green — Blue — Ultraviolet — Frequency →

white when all (RGB) present

UNDER EQUILIBRIUM CONDITIONS, THE RADIATION DEPENDS ONLY ON THE TEMPERATURE. AT ABOUT 800 DEGREES CENTIGRADE, NO MATTER WHAT IS IN THE OVEN — COAL, GLASS, OR METAL — A UNIFORM RED COLOUR IS SEEN.

This is how early potters determined the temperature inside their kilns. Already in 1792, the famous porcelain maker Josiah Wedgwood had noted that all bodies become red at the same temperature.

POTTER'S GUIDE

550 C.	DARK RED
750 C.	CHERRY RED
900 C.	ORANGE
1000 C.	YELLOW
1200 C.	WHITE

In 1896, a friend of Planck's, Wilhelm Wien, and others in the Berlin *Reichsanstalt* (Bureau of Standards) physics department put together an expensive empty cylinder of porcelain and platinum.

WE RECORDED THE *COLOUR DISTRIBUTION* OF RADIATION ALLOWED TO ESCAPE FROM A HOLE IN ONE OF ITS ENDS, MEASURING FROM THE NEAR INFRARED INTO THE VIOLET.

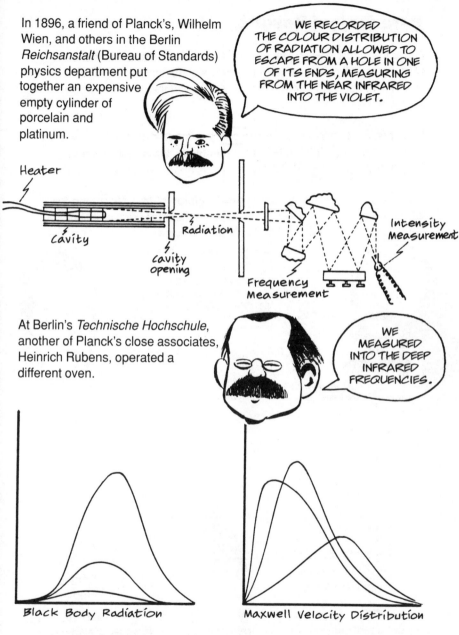

Heater

Cavity

Radiation

Cavity opening

Frequency Measurement

Intensity Measurement

At Berlin's *Technische Hochschule*, another of Planck's close associates, Heinrich Rubens, operated a different oven.

WE MEASURED INTO THE DEEP INFRARED FREQUENCIES.

Black Body Radiation

Maxwell Velocity Distribution

These radiation curves – one of the central problems of theoretical physics in the late 1890s – were shown to be very similar to those calculated by Maxwell for the velocity (i.e. energy) distribution of heated gas molecules in a closed container.

29

Paradoxical Results

Could this *black-body radiation* problem be studied in the same way as Maxwell's ideal gas . . . electromagnetic waves (instead of gas molecules) bouncing around in equilibrium with the walls of a closed container?

Wien derived a formula, based on some dubious theoretical arguments which agreed well with published experiments, but only at the **high frequency** part of the spectrum.

The English classical physicists **Lord Rayleigh** (1842–1919) and **Sir James Jeans** (1877–1946) used the same theoretical assumptions as Maxwell had done with his kinetic theory of gases.

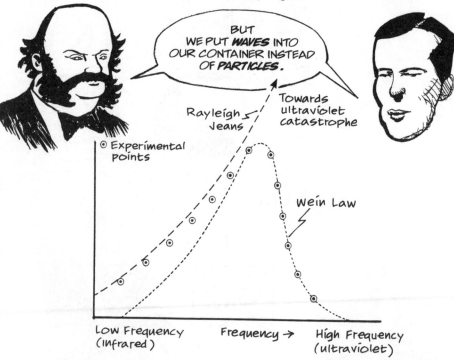

BUT WE PUT *WAVES* INTO OUR CONTAINER INSTEAD OF *PARTICLES*.

Rayleigh Jeans

Towards ultraviolet catastrophe

⊙ Experimental points

Wein Law

Low Frequency (infrared)   Frequency →   High Frequency (ultraviolet)

The equation of Rayleigh and Jeans agreed well at **low frequencies** but they got a real shock at the high frequency region. The classical theory predicted an **infinite intensity for the ultraviolet** region and beyond, as shown in the graph. This was dubbed the ultraviolet catastrophe.

What does this experimental result actually mean?

What Went Wrong?

The Rayleigh-Jeans result is clearly wrong, otherwise anyone who looked into the cavity (or Mr. Wedgwood into his kiln) . . .

I WOULD HAVE MY EYEBALLS BURNED OUT!

THE ULTRAVIOLET CATASTROPHE BECAME A SERIOUS PARADOX FOR CLASSICAL PHYSICS.

IF RAYLEIGH AND JEANS WERE RIGHT, IT WOULD BE DANGEROUS FOR US EVEN TO SIT IN FRONT OF A FIREPLACE.

If classical physicists had their way, the romantic glow of the embers would soon turn into life-threatening radiation. Something had to be done!

The Ultraviolet Catastrophe

Everyone agreed that Rayleigh and Jeans' method was sound, so it is instructive to examine what they actually did and why it didn't work.

WE APPLIED THE STATISTICAL PHYSICS METHOD TO THE *WAVES* BY ANALOGY WITH MAXWELL'S GAS *PARTICLES* USING THE EQUIPARTITION OF ENERGY, I.E. WE ASSUMED THAT THE TOTAL ENERGY OF RADIATION IS DISTRIBUTED EQUALLY AMONG ALL POSSIBLE VIBRATION FREQUENCIES.

BUT THERE IS ONE BIG DIFFERENCE IN THE CASE OF WAVES. THERE IS NO LIMIT ON THE NUMBER OF MODES OF VIBRATION THAT CAN BE EXCITED . . .

Increasing Frequency

½ wave

1 wave

1½ wave

2 waves

etc.

. . . BECAUSE IT'S EASY TO FIT MORE AND MORE WAVES INTO THE CONTAINER AT HIGHER AND HIGHER FREQUENCIES (I.E. THE WAVELENGTHS GET SMALLER AND SMALLER).

CONSEQUENTLY, THE AMOUNT OF RADIATION PREDICTED BY THE THEORY IS UNLIMITED AND SHOULD KEEP GETTING STRONGER AND STRONGER AS THE TEMPERATURE IS RAISED AND THE FREQUENCY INCREASES.

NO WONDER IT WAS KNOWN AS THE *ULTRAVIOLET CATASTROPHE*.

Enter Max Planck

Planck's story begins in the physics department of the Kaiser Wilhelm Institute in Berlin, just before the turn of the century.

> I AM REPEATEDLY BEING CONFRONTED WITH RELIABLE EXPERIMENTAL DATA ON BLACK-BODY RADIATION FROM MY OWN FRIENDS' EXPERIMENTS WHICH SIMPLY CANNOT BE EXPLAINED BY ANY ACCEPTED THEORY.

Planck was a very conservative member of the Prussian Academy, steeped in traditional methods of classical physics and a passionate advocate of thermodynamics. In fact, from his PhD thesis days in 1879 (the year Einstein was born) to his professorship at Berlin twenty years later, he had worked almost exclusively on problems related to the laws of thermodynamics. He believed that the Second Law, concerning entropy, went deeper and said more than was generally accepted.

ANNALEN
DER
PHYSIK.

BEGRÜNDET UND FORTGEFÜHRT DURCH

F. A. C. GREN, L. W. GILBERT, J. C. POGGENDORFF, G. UND E. WIED

VIERTE FOLGE.

BAND 17.

DER GANZEN REIHE 322. BAND.

KURATORIUM

F. KOHLRAUSCH, M. PLANCK, G. QUINC
W. C. RÖNTGEN, E. WARBURG.

UNTER MITWIRKUNG

DER DEUTSCHEN PHYSIKALISCHEN GESELLS

UND INSBESONDERE VON

PLANCK

Planck was attracted by the absolute and universal aspects of the black-body problem. Plausible arguments showed that at equilibrium, the curve of radiation intensity versus frequency should not depend on the size or shape of the cavity or on the materials of its walls. The formula should contain only the temperature, the radiation frequency and one or more universal constants which would be the same for all cavities and colours.

Finding this formula would mean discovering a relationship of quite fundamental theoretical interest.

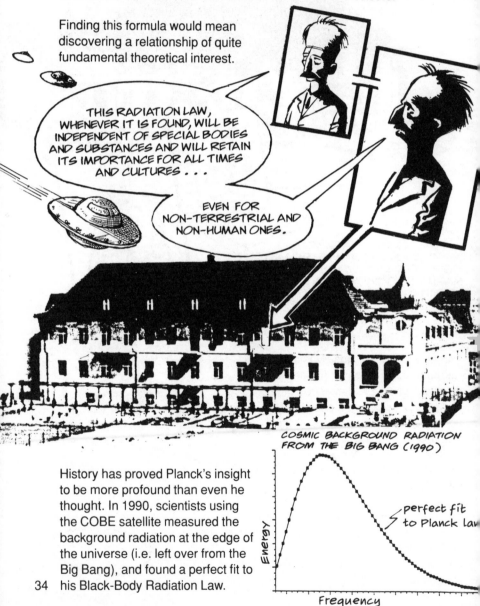

THIS RADIATION LAW, WHENEVER IT IS FOUND, WILL BE INDEPENDENT OF SPECIAL BODIES AND SUBSTANCES AND WILL RETAIN ITS IMPORTANCE FOR ALL TIMES AND CULTURES . . .

EVEN FOR NON-TERRESTRIAL AND NON-HUMAN ONES.

COSMIC BACKGROUND RADIATION FROM THE BIG BANG (1990)

perfect fit to Planck Law

Energy

Frequency

History has proved Planck's insight to be more profound than even he thought. In 1990, scientists using the COBE satellite measured the background radiation at the edge of the universe (i.e. left over from the Big Bang), and found a perfect fit to his Black-Body Radiation Law.

34

Pre-Atomic Model of Matter

Planck knew the measurements by his friends Heinrich Rubens and
Ferdinand Kurlbaum were extremely reliable.

Experimental cavity

Planck's oscillators in the walls of the cavity

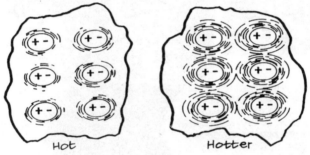

Hot   Hotter

Planck started by introducing the idea of a collection of electric oscillators*
in the walls of the cavity, vibrating back and forth under thermal agitation.
(*Note! Nothing was known about atoms.)
Planck assumed that all possible frequencies would be present. He also
expected the **average** frequency to increase at higher temperatures as
heating the walls caused the oscillators to vibrate faster and faster until
thermal equilibrium was reached.

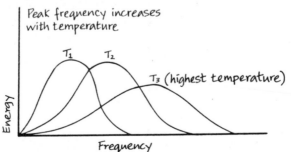

The electromagnetic theory could tell everything about the emission, absorption and propagation of the radiation, but nothing about the energy distribution at equilibrium. This was a thermodynamic problem.

Planck made certain assumptions, relating the average energy of the oscillators to their entropy, thereby obtaining a formula for the intensity of the radiation which he hoped would agree with the experimental results.

INITIALLY, I SIMPLY WANTED TO FIND A PROOF FOR THE FORMULA DERIVED BY MY FRIEND *WIEN*, WHICH EVERYONE THOUGHT TO BE CORRECT.

BUT THE MOST RECENT MEASUREMENTS BEGAN TO CAST DOUBTS ON WIEN'S EQUATION IN THE *INFRARED OR LOW FREQUENCY REGION*. DISCREPANCIES GREATER THAN THE POSSIBLE EXPERIMENTAL ERRORS WERE FOUND.

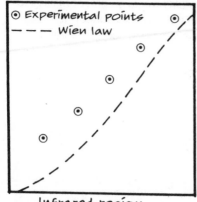

⊙ Experimental points
--- Wien law

Infrared region

Planck tried to alter his expression for the entropy of the radiation by generalizing it, and eventually arrived at a new formula for the radiation intensity over the entire frequency range.

The constants C_1 and C_2 are numbers chosen by Planck to make the equation fit the experiments.

Among those present at the historic seminar was Heinrich Rubens. He went home immediately to compare his measurements with Planck's formula. Working through the night, he found perfect agreement and told Planck early next morning.

Planck had found the correct formula for the radiation law. Fine. But could he now use the formula to discover the underlying physics?

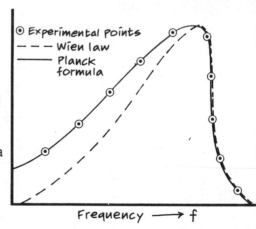

Planck's Predicament

. . . FROM THE VERY DAY I FORMULATED THE RADIATION LAW, I BEGAN TO DEVOTE MYSELF TO THE TASK OF INVESTING IT WITH TRUE PHYSICAL MEANING.

AFTER TRYING EVERY POSSIBLE APPROACH USING TRADITIONAL CLASSICAL APPLICATIONS OF THE LAWS OF THERMODYNAMICS, I WAS DESPERATE.

COME ON *MAX*, DON'T BE SO STUBBORN, IT'S WORTH A TRY.

Boltzmann's statistical version of the Second Law based on probabilities seemed Planck's only alternative. But he rejected the underlying assumption of Boltzmann's approach which allows the Second Law to be violated momentarily during fluctuations.

I WAS FORCED TO CONSIDER THE RELATION BETWEEN ENTROPY AND PROBABILITY ACCORDING TO *BOLTZMANN'S* IDEAS. AFTER SOME OF THE MOST INTENSE WEEKS OF MY LIFE, THE LIGHT BEGAN TO APPEAR TO ME . . .

That light was

$$S = k \, Log \, W$$

(**Boltzmann's version of the Second Law of Thermodynamics**).

Not once in any of the forty or so papers that Planck wrote prior to 1900 did he use, or even refer to, Boltzmann's statistical formulation of the Second Law!

Chopping Up the Energy

So, Planck applied three of Boltzmann's ideas about entropy.
1) His statistical equation to **calculate** the entropy.
2) His condition that the entropy must be a maximum (i.e. totally disordered) at equilibrium.
3) His counting technique to determine the probability **W** in the entropy equation.

To calculate the probability of the various possible arrangements, Planck followed Boltzmann's method of dividing the energy of the oscillators into arbitrarily small but **finite** chunks. So the total energy was written as $E = N e$ where **N** is an integer and **e** an arbitrarily small amount of energy. **e** would eventually become infinitesimally small as the chunks became infinite in number, consistent with the mathematical procedure.

A Quantum of Energy

Eureka! Planck had stumbled across a mathematical method which at last gave some theoretical basis for his experimental radiation law – **but only if the energy is discontinuous**.

Even though he had no reason whatsoever to propose such a notion, he accepted it provisionally, for he had nothing better. He was thus forced to postulate that the quantity **e = h f** must be a finite amount and **h** is not zero.

Thus, if this is correct, it must be concluded that it is not possible for an oscillator to absorb and emit energy in a continuous range. It must gain and lose energy discontinuously, in small indivisible units of **e = h f**, which Planck called "energy quanta".

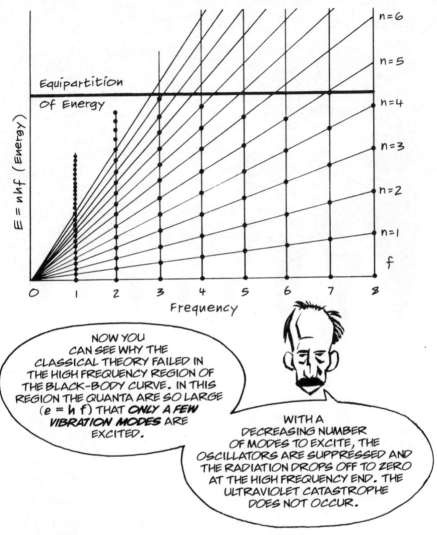

Planck's quantum relation thus inhibits the equipartition of energy and not all modes have the same total energy. This is why we don't get sunburn from a cup of coffee. (Think about it!)

The classical approach of Rayleigh-Jeans works fine at low frequencies, where **all** the available vibrational modes can be excited. At high frequencies, even though plenty of modes of vibration are **possible** (recall it's easier to stuff **short** waves into a box), not many are excited because it costs too much energy to make a quantum at a high frequency, since **e = h f**.

41

During his early morning walk on 14 December 1900, Planck told his son that he may have produced a work as important as that of Newton. Later that same day, he presented his results to the Berlin Physical Society signalling the birth of quantum physics.

It had taken him less than two months to find an explanation for his own black-body radiation formula. Ironically, the discovery was accidental, caused by an incomplete mathematical procedure. **An ignominious start to one of the greatest revolutions in the history of physics!**

From this start would come an understanding of why statistical rules must be used for atoms, why atoms don't glow all the time and why atomic electrons don't spiral into the nucleus.

In early 1901, the constant **h** – today called Planck's constant – appeared in print for the first time. The number is small –

h = 0.000 000 000 000 000 000 000 000 006 626

–**but it is not zero!** If it were, we would never be able to sit in front of a fire. In fact, the whole universe would be different. *Be thankful for the little things in life.*

Surprisingly, in spite of the important and revolutionary aspects of the black-body formula, it did not draw much attention in the early years of the 20th century. Even more surprisingly, Planck himself was not convinced of its validity.

I WAS SO SCEPTICAL OF THE UNIVERSALITY OF BOLTZMANN'S ENTROPY LAW THAT I SPENT YEARS TRYING TO EXPLAIN MY RESULTS IN A LESS REVOLUTIONARY WAY.

NEVERTHELESS, QUANTUM THEORY HAD BEEN BORN.

Now to the second experiment which could not be explained by classical physics. It is more simple, yet inspired a more profound explanation.

43

The Photoelectric Effect

While Max Planck was struggling with the black-body problem, another German physicist, **Philipp Lenard** (1862–1947) was focusing beams of *cathode rays* (soon to be identified as electrons) at thin metal foils.

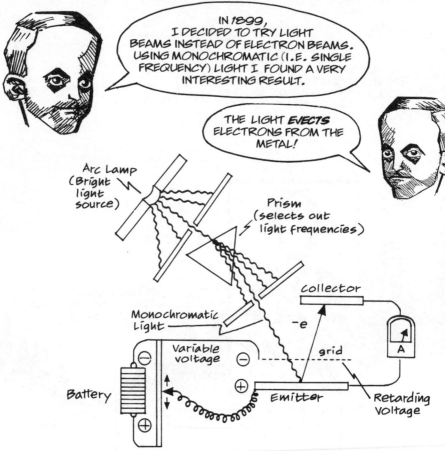

IN *1899*, I DECIDED TO TRY LIGHT BEAMS INSTEAD OF ELECTRON BEAMS. USING MONOCHROMATIC (I.E. SINGLE FREQUENCY) LIGHT I FOUND A VERY INTERESTING RESULT.

THE LIGHT *EVECTS* ELECTRONS FROM THE METAL!

Arc Lamp (Bright light source)

Prism (selects out light frequencies)

Collector

Monochromatic Light

-e

grid

A

Variable voltage

Battery

Emitter

Retarding Voltage

Though this effect had been noticed by Heinrich Hertz ten years before, Lenard was now able to measure some properties of these *photoelectrons* with a simple electrical circuit.

The ejected electrons are produced by illumination of the metal plate called the *emitter* and received at another plate called the *collector*. The total photoelectric current is measured on the sensitive current-measuring device marked A. The electrical potential or *voltage* between emitter and collector can be varied and has a strong effect on the measured current.

The current decreases sharply when a **retarding** voltage is applied, making the collecting electrode negative with respect to the emitting electrode. (Electrons have negative charge and are repelled by a negative voltage.) At a definite value of the retarding voltage, marked as V_0 in the diagram, the photoelectric current disappears entirely.

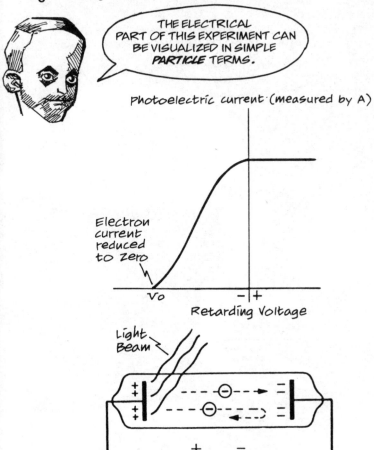

THE ELECTRICAL PART OF THIS EXPERIMENT CAN BE VISUALIZED IN SIMPLE *PARTICLE* TERMS.

Photoelectric current (measured by A)

Electron current reduced to zero

V_0

Retarding Voltage

Light Beam

V_0

Ejected electrons leave the target plate with a certain kinetic energy and continuously lose this energy as they travel against the retarding negative voltage between the emitter and collector plates.

Electrons which are collected and contribute to the measured current must have had (when initially emitted) at least the energy greater than qV_0 (q is the charge of the electron). This is the well-known relationship for the energy of an electron under the action of a voltage.

45

A Classical Interpretation

A straightforward interpretation would conclude that the emitted electrons must acquire their kinetic energy from the **light beam** shining on the metal surface.

The classical viewpoint would assume that the light waves beat on the metal surface like ocean waves and the electrons are disturbed like pebbles on a beach. Clearly, **more intense illumination** (i.e. brighter) would deliver **more energy to the electrons**.

THIS IS NOT WHAT I FOUND. IN 1902, I DISCOVERED THAT THE ELECTRON ENERGIES — AS MEASURED BY THE RETARDING POTENTIALS — WERE ENTIRELY INDEPENDENT OF THE LIGHT INTENSITY.

Further experiments showed another unexplained effect. There was a certain **threshold** frequency below which **no photoelectrons** were ejected, no matter how bright the light beam. This was very strange indeed. Real trouble here for the classical guys.

Enter Albert Einstein

This time it was not an established, respected university professor who solved the problem, but a young clerk at the Swiss Patent Office in Bern.

In 1905, at the tender age of 26, Einstein published three papers in a single volume of *Annalen der Physik*.

NOBEL PRIZE

THESE WERE THE *LIGHT QUANTUM PAPER*, WHICH CONCERNS US NOW; A SECOND SEMINAL PAPER WHICH PROVED THE *EXISTENCE OF ATOMS*; AND A THIRD WHICH INTRODUCED *RELATIVITY*, SOLVING SERIOUS PROBLEMS IN ELECTRODYNAMICS AND MOTION.

Einstein was familiar with the experimental puzzles of the photoelectric experiment and knew of the work of Planck and his radiation law. Yet his approach was utterly personal, relying on his own statistical approach to physics and Boltzmann's expression for the entropy of a collection of particles.

47

Einstein with his wife Mileva (a trained engineer) and young son Hans Albert . . .

MILEVA LIEBCHEN, I WANT TO SHOW YOU MY LATEST CALCULATION. I THINK IT MAY BE QUITE PROFOUND. FIRST OF ALL, YOU REMEMBER THE IMPORTANT LAW OF BOLTZMANN FOR THE ENTROPY OF A SYSTEM OF PARTICLES IN TERMS OF ITS PROBABILITY, $S = k \log W$ . . .

$$S = k \log W \ldots$$

OH YES, THE EQUATION THAT *PLANCK* HATED BUT WAS FORCED TO USE ON THE BLACK-BODY PROBLEM.

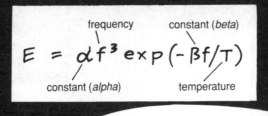

ALSO, YOU MAY REMEMBER THE RADIATION FORMULA OF WILHELM WIEN, PLANCK'S COLLEAGUE IN BERLIN, WHICH EVERYONE AGREED WAS VALID FOR THE HIGH FREQUENCY PORTION OF THE BLACK-BODY CURVE.

YES, I DO. IN FACT, DIDN'T PLANCK'S RADIATION FORMULA REDUCE TO WIEN'S FOR HIGH FREQUENCIES?

VERY GOOD, MILEVA.

BUT I DON'T WANT TO USE PLANCK'S THEORETICAL FORMULA.

I WOULD RATHER BASE MY WORK ON WIEN'S EMPIRICAL LAW WHICH WE KNOW FITS THE HIGH FREQUENCY EXPERIMENTS SO WELL. I'M USING A PHENOMENOLOGICAL METHOD RATHER THAN A STRICTLY THEORETICAL ONE.

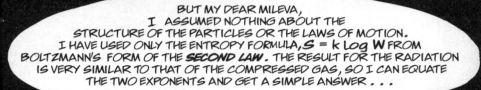

BUT MY DEAR MILEVA,
I ASSUMED NOTHING ABOUT THE
STRUCTURE OF THE PARTICLES OR THE LAWS OF MOTION.
I HAVE USED ONLY THE ENTROPY FORMULA, $S = k \log W$ FROM
BOLTZMANN'S FORM OF THE **SECOND LAW**. THE RESULT FOR THE RADIATION
IS VERY SIMILAR TO THAT OF THE COMPRESSED GAS, SO I CAN EQUATE
THE TWO EXPONENTS AND GET A SIMPLE ANSWER . . .

$$E = n k \beta f$$

SO MY HYPOTHESIS IS THIS . . .
WITHIN THE VALIDITY OF THE **WIEN LAW**
(I.E. HIGH FREQUENCY), RADIATION BEHAVES
THERMODYNAMICALLY AS IF IT CONSISTS OF
MUTUALLY INDEPENDENT ENERGY QUANTA OF
MAGNITUDE $k \beta f$. IN OTHER WORDS, LIKE
LIGHT PARTICLES.

ONE MORE THING, ALBERT. I NOTICE YOU HAVE THE CONSTANT β FROM THE WIEN LAW IN YOUR ANSWER. BUT DIDN'T PLANCK SHOW THAT β CAN BE WRITTEN AS A RATIO OF HIS OWN CONSTANT h AND THE BOLTZMANN CONSTANT, k?

YES. BUT I DIDN'T WANT TO INCLUDE ANY CONCLUSIONS OF THE PLANCK RADIATION LAW IN MY PAPER BECAUSE I AM NOT CERTAIN OF ITS RESULTS. I AM SUGGESTING THE QUANTIZATION OF *ALL* LIGHT RADIATION.

PLANCK MERELY CONSIDERED OSCILLATORS IN THE CAVITY WALLS.

WHAT HAPPENS IF YOU ELIMINATE WIEN'S β?

WELL . . . $\beta = h/k$ SO, $E = n h f$

IF I DO THAT, I GET AN EQUATION FOR THE ENERGY OF THE RADIATION WHICH IS EQUAL TO THE *NUMBER OF PARTICLES TIMES THE QUANTITY* $h f$, WHICH CLEARLY INDICATES THAT $h f$ IS THE QUANTUM OF RADIATION.

THIS WOULD MEAN THAT *ALL LIGHT AND ELECTROMAGNETIC RADIATION* TRAVELS IN BUNDLES OF ENERGY EQUAL TO $h f$. IT IS A MUCH MORE GENERAL RULE THAN ANYTHING PLANCK EVER IMAGINED!

Einstein's Explanation of the Photoelectric Effect

Einstein's 1905 paper showed that the puzzling features of the photoelectric effect are easily explained once the illuminating radiation is understood to be a collection of particles or photons. If the photons can transfer their energy to the electrons in the target metal, a complete and simple picture is possible. Let's see how this works . . .

IF ONE ASSUMES THAT THE INCIDENT LIGHT CONSISTS OF ENERGY QUANTA (PHOTONS) OF MAGNITUDE hf, IT IS POSSIBLE TO CONCEIVE OF THE EJECTION OF ELECTRONS BY LIGHT AS FOLLOWS. ENERGY QUANTA PENETRATE THE SURFACE LAYER OF THE METAL OF THE TARGET ELECTRODE. THEIR ENERGY IS TRANSFERRED, AT LEAST IN PART, INTO THE KINETIC ENERGY OF THE ELECTRONS AND SOME ARE EJECTED.

The simplest way to imagine this is to assume that a light quantum delivers its entire energy – hf – to the electron which then loses some of this energy by the time it reaches the surface.

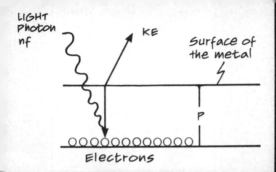

LIGHT
Photon
nf

KE

Surface of
the metal

P

Electrons

...ng ejected, each electron must perform an amount of work – ...teristic of the metal to get out into free space. The electrons ...e metal with the **largest** velocity will be those located near the ...hich will have minimum work to get free. The kinetic energy of ...ns is given by . . .

> KINETIC ENERGY = hf
> (ENERGY OF THE INCOMING PHOTON)
> LESS P (WORK TO GET OUT
> OF THE METAL).

If the plate must be charged to a voltage V_o in order to offset the kinetic energy and reduce the electron current to zero, (i.e. stopping even the most energetic ejected electrons), it follows that

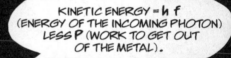

$$qV_o = hf - P,$$

WHERE q DENOTES THE
ELECTRON CHARGE.

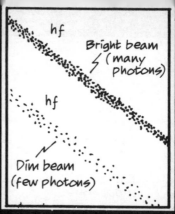

hf

Bright beam
(many
photons)

hf

Dim beam
(few photons)

derived a very simple equation for the photoelectro
ed in the laboratory. Furthermore, since each intera
ame photon-electron energy transfer, the observat
gies do not respond to changes in light intensity wa
te simply. The intensity affects the **number** of phot
magnitude of the electron current, but does not affe
e V_O which is determined by frequency.

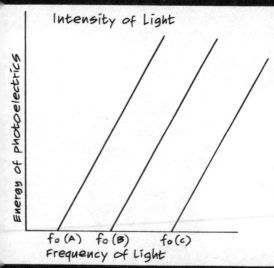

Intensity of Light

Energy of photoelectrics

f_o (A) f_o (B) f_o (C)
Frequency of Light

consequence of these arguments and simple equa
imum retarding potential V_O is a linear function of t
the incident light. Thus in time-honoured fashion, i
ht-line) relationship could be tested, it would provic
f Einstein's **photon** concept. The experiments mus
(the cut-off voltage) for several different light frequ

Millikan: Hard-headed Classical Physicist

During the years 1912–17, **Robert A. Millikan** (1868–1953), working in the Ryerson Laboratories at the University of Chicago, submitted the Einstein equation to this linearity test. He used several different metals, including highly-reactive sodium, as targets and illuminated them with light of various frequencies.

His technique was impeccable, even scraping the surface of the metals in vacuum to avoid oxidized layers which might affect the results. He **always obtained linear results** . . . and yet was very disappointed.

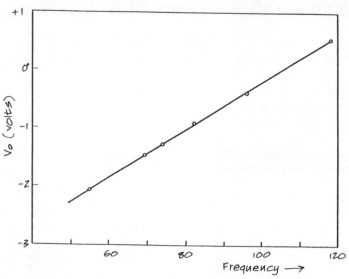

Millikan's results

Yet Millikan only strengthened Einstein's explanation by obtaining remarkably accurate data with near perfect linearity. In fact, it eventually won him a Nobel Prize.

CONTRARY TO ALL MY EXPECTATIONS, I AM COMPELLED TO ASSERT ITS UNAMBIGUOUS EXPERIMENTAL VERIFICATION IN SPITE OF ITS UNREASONABLENESS.

THE HYPOTHESIS WAS MADE SOLELY BECAUSE IT FURNISHED A READY EXPLANATION OF THE FACT THAT THE ENERGY OF AN EJECTED ELECTRON IS INDEPENDENT OF THE INTENSITY OF THE LIGHT . . . BUT DEPENDS ON THE FREQUENCY. I UNDERSTAND EVEN EINSTEIN HIMSELF NO LONGER ACCEPTS IT.

Such sentiments were typical of physicists in the second decade of the 20th century. Clearly, the prediction of quantized radiation was **not** a great triumph for Planck and Einstein.

58

IN FACT, DURING THIS PERIOD OUR WORK WAS COMPLETELY IGNORED.

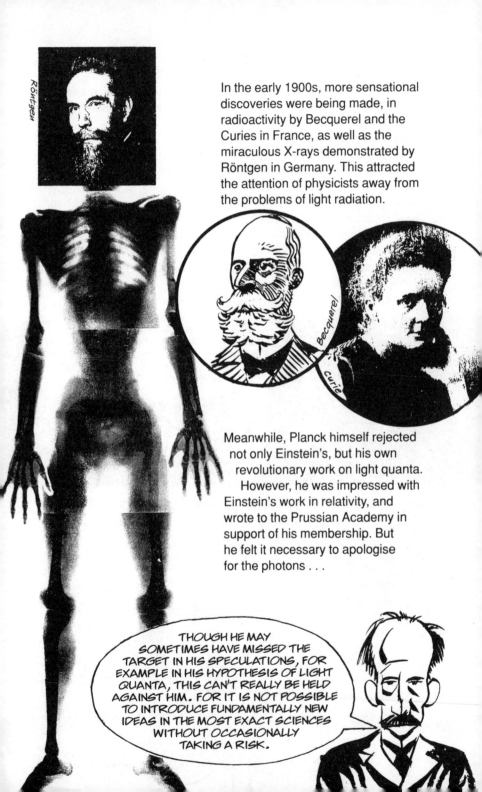

Röntgen

In the early 1900s, more sensational discoveries were being made, in radioactivity by Becquerel and the Curies in France, as well as the miraculous X-rays demonstrated by Röntgen in Germany. This attracted the attention of physicists away from the problems of light radiation.

Becquerel

curie

Meanwhile, Planck himself rejected not only Einstein's, but his own revolutionary work on light quanta. However, he was impressed with Einstein's work in relativity, and wrote to the Prussian Academy in support of his membership. But he felt it necessary to apologise for the photons . . .

THOUGH HE MAY SOMETIMES HAVE MISSED THE TARGET IN HIS SPECULATIONS, FOR EXAMPLE IN HIS HYPOTHESIS OF LIGHT QUANTA, THIS CAN'T REALLY BE HELD AGAINST HIM. FOR IT IS NOT POSSIBLE TO INTRODUCE FUNDAMENTALLY NEW IDEAS IN THE MOST EXACT SCIENCES WITHOUT OCCASIONALLY TAKING A RISK.

Bright Line Light Spectra

We are now ready for the third experiment which could not be explained by the classical physicists – Bright Line Light Spectra. Remember the list...

Black-Body Radiation (explained by Planck)
The Photoelectric Effect (explained by Einstein)
Bright Line Light Spectra (to be explained by Bohr)

For 150 years, precise observations of light emission from gases had been accumulating in European physics laboratories. Many believed these held the secrets of the atom. But how to decipher this vast store of information to create order from chaos? That was the challenge. Reports began as far back as 1752 when the Scottish physicist, Thomas Melvill, put containers of different gases over a flame and studied the glowing light emitted.

AFTER PLACING A PASTEBOARD WITH A CIRCULAR HOLE IN IT BETWEEN MY EYE AND THE FLAME . . . I EXAMINED THE CONSTITUTION OF THESE DIFFERENT LIGHTS WITH A PRISM.

Melvill made a rather remarkable discovery. He found that the *spectrum* of light from a hot gas when passed through a prism was completely different from the well known rainbow-like spectrum of a glowing solid.

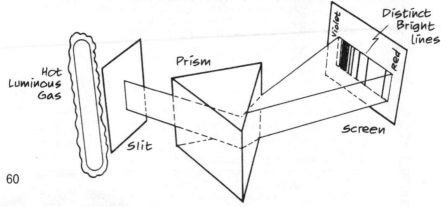

BRIGHT LINE SPECTRUM OF A LUMINOUS GAS

Emission Spectra

When examined through a narrow slit, the light spectrum from a heated gas consists of **distinct bright lines,** each having the colour of the part of the spectrum in which it was located. Different gases gave different patterns.

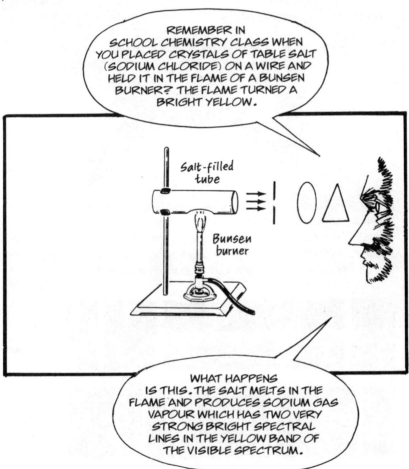

The integrating properties of the eye keep us (and other animals) from seeing these lines, combining the separate colours so that we see only the mixture (e.g. reddish for glowing neon, pale blue for nitrogen, and so on). In the case of sodium, the eye mixes the two yellow lines and the flame looks like fiery daffodil petals.

Mercury gas (from vaporized liquid) and **nitrogen** gas gave sharply defined and easily recognizable bright line patterns when photographed with a sensitive device called a *spectrometer*.

In fact, the spectral patterns of elements are so distinct and the measurements so exact, that no two are known to have the same set of lines. Spectra could be used to identify **unknown** gases, as with the discovery of helium gas in the spectrum of the sun. But before describing this amazing discovery, a word about **dark lines** in light spectra.

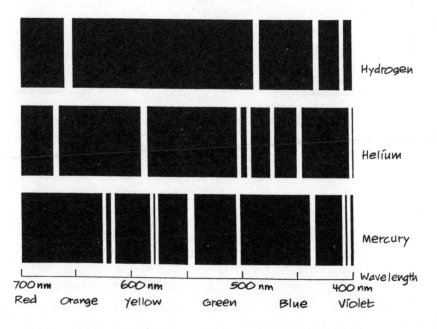

Hydrogen

Helium

Mercury

Wavelength

700 nm	600 nm	500 nm	400 nm
Red	Orange	Yellow	Green	Blue	Violet

Absorption Spectra (Dark Lines)

These three diagrams show how to observe the two different types of spectra.

1) "White light" radiation containing all frequencies is emitted from a hot solid (like the heated filament in a light bulb) and passes through a slit before entering the thin end of a triangular prism wedge. A **continuous spectrum** (rainbow-like) appears on the screen.

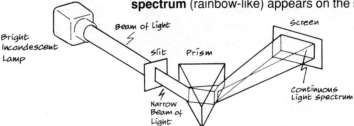

2) The same experimental set-up is used, except a **hot gas** is used as a source in place of a **hot solid**. Now a **bright line spectrum** appears on the screen and the shape of each line is the image of the slit.

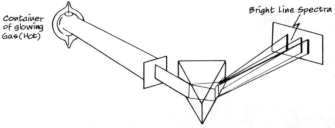

3) Now, **something new**. Return to the first case, the hot solid giving off radiation of all frequencies. The container of gas is interposed between the source and the slit. But this time, the gas is not heated . . . it is cool.

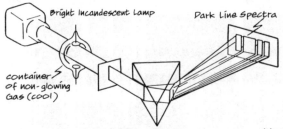

Now note the screen. **A dark line spectrum** appears with the missing lines corresponding exactly to the **bright lines** of the previous case, when the gas was hot.

A simple conclusion can be drawn. The cool (unexcited) gas is *absorbing* light at precisely the same frequencies at which this same gas *emitted* light when heated. There must be certain characteristic energy states in a gas which are reversible, i.e. can *take in* or *give off* energy. Very interesting . . .

Fraunhofer Lines

All this was very puzzling, but at the same time encouraging because in both the emission and absorption spectra, the frequency (or wavelength) at which these lines appeared was always the same. Line spectra gave physicists precise reproducible information about pure elements.

In 1814, **Joseph von Fraunhofer** (1787–1826) created the first *spectroscope*, combining a prism with a small viewing telescope focused on a distant narrow slit. He subsequently used the instrument to view the sun's spectrum and saw . . .

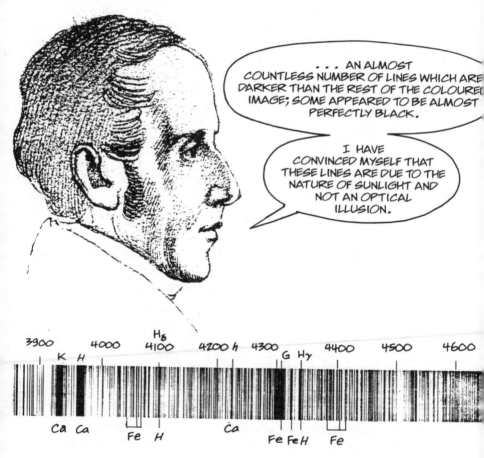

These dark lines in the solar spectrum became known as *Fraunhofer Lines* and form the basis for astrophysical spectroscopy.

The Discovery of Helium

Gustav Kirchhoff (1824–87) studied these dark lines several years later using an ingenious method of superimposing the bright yellow lines from a salt (NaCl) solution onto the Fraunhofer solar spectrum. The exact match demonstrated that the dark lines were due to the presence of cool vapours of sodium and other elements in the outer atmosphere surrounding the sun.

The elusive element – an odourless, colourless and chemically inert gas – was finally detected and isolated. Appropriately, it was named **helium**, after the Greek word (*helios*) for the sun.

65

Hydrogen – Test Case for Atomic Structure

Surely these line spectra must be revealing something quite fundamental about the internal structure of the atom. But what? A closer examination was called for.

In attempts to relate the characteristic bright lines to some kind of theory of atomic structure, it is not surprising that physicists chose to examine the spectra of **hydrogen**. It is the simplest atom of all the elements.

The four most prominent lines of hydrogen, all in the visible part of the light spectrum, had been measured accurately as early as 1862 by the Swedish astronomer, **A. J. Ångstrom** (1814–74).

Balmer: the Swiss School Teacher

In 1885, a Swiss school mathematics teacher, **Johann Jakob Balmer** (1825–98), published the results of months of work spent manipulating the numerical values of the frequencies of the lines of the visible hydrogen spectrum.

I WAS SIMPLY PROVIDING AN INITIAL ORGANIZATION OF THE RAW DATA. NO REAL PHYSICS WAS INVOLVED, IT WAS PURE NUMEROLOGY.

Miraculously, Balmer had managed to discover a formula involving whole numbers which **predicted almost exactly** the frequencies of the four visible hydrogen lines – and others in the ultraviolet region, later confirmed.

Rydberg constant

$$f = R \left(\frac{1}{n_f^2} - \frac{1}{n_i^2} \right)$$

Using this equation, Balmer could predict the frequencies of the four hydrogen lines if n_f (final) was chosen to be 2; n_i (initial) = 3, 4, 5 and 6 and **R** was given the value 3.29163×10^{15} cycles/sec. This gave the best fit to the measurements.

67

A comparison of Balmer's values with the actual measurements is shown in the table below.

Hydrogen Emission Spectrum (Balmer, 1885)

Experimental Values		From Balmer Formula	
Wavelength	Frequency	Frequency	Value of n_i
$(nm = 10^{-9} m)$	$(10^6 Mhz)$	$(10^6 Mhz)$	$(n_f = 2)$
656.210 (red)	457.170	457.171	3
486.074 (green)	617.190	617.181	4
434.01 (blue)	691.228	691.242	5
410.12 (violet)	731.493	731.473	6

LOOK AT THE EXTREMELY CLOSE AGREEMENT BETWEEN THE EXPERIMENTAL FREQUENCIES AND THE FREQUENCIES I COMPUTED USING MY EQUATION.

The accuracy was too good to be *not true*! There must be something fundamental underlying his equation. Perhaps certain physical laws applied to the atom might generate an equation of this form.

Meanwhile, Balmer predicted more lines – in the ultraviolet and infrared frequency range – which could not even be measured at the time. He used different values for n_f and predicted several series of spectra lines.

Balmer's equation predicted an infinity of lines . . . and was, as you shall see quite correct! But whether it would lead to a new theory remained to be seen.

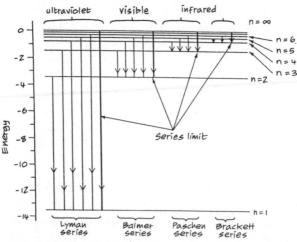

The longer the length of the arrow, the higher the frequency of the light.
The origins of the various series of spectral lines are indicated on the diagram.

Hydrogen Frequencies From Balmer's Formula

Balmer speculated that more hydrogen lines exist with n_f taking on values other than 2. For example, $n_f = 1$ gave a new series in the ultraviolet and $n_f = 3$ and 4 gave other new series in the infrared.

Table of Hydrogen Spectral Series (Balmer's Equation)

n_{final}	$n_f = 1$	$n_f = 2$	$n_f = 3$	$n_f = 4$
$n_{initial}$	$n_i = 2,3,4,5,6\ldots$	$n_i = 3,4,5,6,7\ldots$	$n_i = 4,5,6,7,8\ldots$	$n_i = 5,6,7,8,9\ldots$
light band	ultraviolet	visible	infrared	infrared
discovery	1906–14	1885	1908	1922
discoverer	Lyman	Balmer	Paschen	Brackett

These sequences strongly suggested some kind of energy diagram as the **emission/absorption** of light from an atom must correspond to a **decrease/increase** in the atom's energy. The diagram below shows how Balmer's formula was used to predict the frequencies of the spectral lines by starting each sequence with a different number, as in the table.

Hydrogen Frequencies From Balmer's Formula

This information would be critical to any atomic theory. These changes of **whole numbers**, which gave precise frequencies of the emitted radiation, suggested some rearrangement of the parts of the atom was taking place.

No one had any idea of the make-up of an atom in the 1890s. Yet it seemed clear that a successful theory of the atom must include the miraculous formula of Johann Jakob Balmer in some significant way.

Discovery of the Electron

It was in the hallowed halls of the world-famous Cavendish Laboratory of Cambridge University that the atom began to be dissected by **J.J. Thomson** (1856–1940), one of the great classical physicists of the 19th century.

I DEMONSTRATED THAT THE ELECTRON HAD A DISTINCT CHARGE-TO-MASS RATIO AND WAS THUS A *PARTICLE*, NOT A CATHODE *RAY*.

In fact, during the last five years of the 19th century, other so-called *rays* were shown to behave as *particles*. Alpha and beta rays became alpha and beta particles. The next step was to see how these particles might be used to make an atom.

Christmas Pudding Atom

Thomson and Lord Kelvin developed a model of the atom (probably at Christmas time) in which the negative electrons were embedded in a uniform sphere of positive charge, like raisins in a pudding. The usual classical assumptions were to apply:

RADIATION FROM THE ATOM SHOULD BE DESCRIBED BY MAXWELL'S ELECTROMAGNETIC THEORY.

THE DYNAMICS IN THE ATOM SHOULD FOLLOW NEWTON'S LAWS OF MOTION.

Raisin Cake model of The Atom. The negativity charged Electrons (the raisins) are embedded in a uniform sphere of positive charge (the cake).

Though well-publicized, this scheme was inherently unstable and got nowhere.

Then, about 1907, one of the more imaginative, perhaps even iconoclastic, of the classical physicists moved to centre stage. **Ernest Rutherford** (1871–1937), a former student of Thomson's at Cambridge, was by this time professor of physics at the University of Manchester and working in the new research area of radioactivity.

Rutherford's Nuclear Atom

Though at heart an ardent experimentalist, Rutherford was always ready to work on a theoretical model if based on reliable measurements which he could see and understand.

He worked closely with his research students, encouraging them regularly by strolling through the laboratories singing "Onward Christian Soldiers".

RUTHERFORD'S ALPHA-
SCATTERING SET-UP

In 1908, while continuing a programme of research on the radioactive alpha particles, Rutherford got the idea that these massive, positively-charged projectiles might be the ideal probes to study the structure of the atom. With one of his students from Germany, **Hans Geiger** (1882–1945), Rutherford began to study alpha-scattering by a thin gold foil, observing through a microscope the tiny flashes produced as individual alpha particles struck a fluorescent screen.

The next day . . .

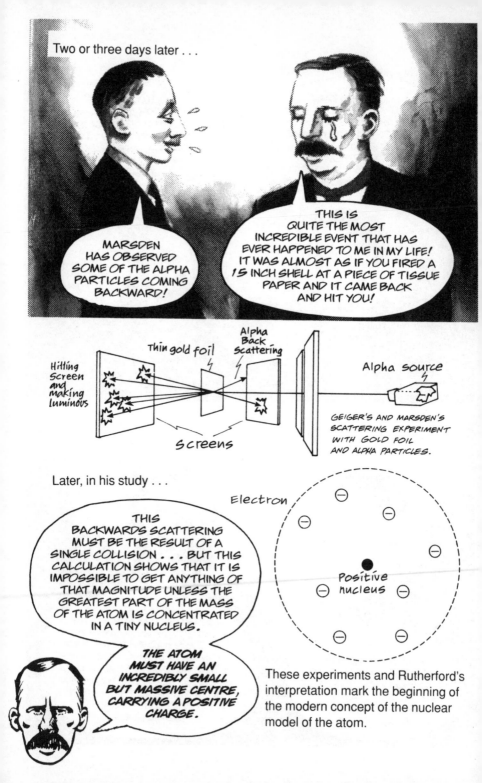

These experiments and Rutherford's interpretation mark the beginning of the modern concept of the nuclear model of the atom.

Size of the Nucleus

As a secondary result of these scattering experiments, the size of the nucleus could be estimated. If an alpha particle moves **directly** towards a nucleus, its kinetic energy on approach is transformed to electrical potential energy until it slows down and eventually stops. The distance of closest approach can then be computed from the conservation of energy.

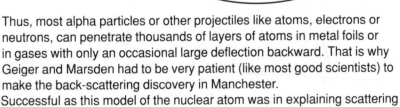

THE ATOM IS MOSTLY EMPTY, WITH THE NUCLEUS OCCUPYING ABOUT ONLY ONE BILLIONTH OF THE SPACE!

Thus, most alpha particles or other projectiles like atoms, electrons or neutrons, can penetrate thousands of layers of atoms in metal foils or in gases with only an occasional large deflection backward. That is why Geiger and Marsden had to be very patient (like most good scientists) to make the back-scattering discovery in Manchester.

Successful as this model of the nuclear atom was in explaining scattering phenomena, it raised many new questions.

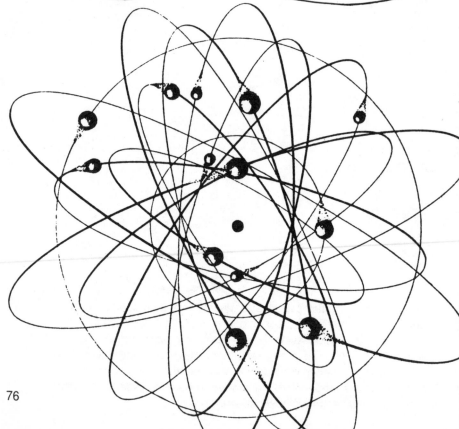

Yet this presented him with yet another problem . . .

IF THE ELECTRONS ARE LIKE A MICROSCOPIC SOLAR SYSTEM, MOVING IN CIRCULAR ORBITS ABOUT THE NUCLEUS (AND THUS ACCELERATING) WHAT KEEPS THEM FROM RADIATING CONTINUOUSLY AS THE CLASSICAL ELECTRO-MAGNETIC THEORY WOULD PREDICT?

THEY WOULD LOSE ALL THEIR ENERGY IN A FRACTION OF A SECOND.

Rutherford's model was unstable.

ONE SHOULD NOT EXPECT A MODEL – MADE ON THE BASIS OF ONE SET OF PUZZLING RESULTS WHICH IT HANDLED WELL – ALSO TO HANDLE ALL OTHER PUZZLES.

ADDITIONAL ASSUMPTIONS WOULD BE NEEDED TO COMPLETE THE PICTURE, PARTICULARLY WITH REGARD TO THE DETAILS OF ATOMIC STRUCTURE.

At least the visualization of the atom had begun. The next step also took place in Rutherford's group at Manchester, with the arrival of a young Danish student, recently transferred from Cambridge . . .

Arrival of the Quantum Hero, Niels Bohr

At Rutherford's Manchester laboratory in 1912, the "Great Dane" began his relentless search for the deepest understanding of quantum physics, which continued for 50 years to his death in 1962.

In this great endeavour, there is no one to compare with Bohr, not even Einstein. He is the grandfather of quantum physics, proposing the first ideas and working with just about everyone who made contributions to the theory's development.

He arrived in England in 1911 with a large dictionary and the complete works of Charles Dickens from which to study English. In spite of his language limitations, Bohr had great self-confidence and an unbelievable capacity for hard work.

Then Bohr met Rutherford at a Cavendish dinner and was very impressed with the enthusiasm and praise Rutherford had expressed for the work of **someone else**.

When Bohr arrived at Manchester, the place was buzzing with the application of Rutherford's new planetary atom. He was not intimidated by the restrictions on Rutherford's model and felt intuitively that classical mechanics did not apply inside the atom anyway. He knew that the work of Planck and Einstein on light radiation was very important, not just a clever German idea.

As early as the summer of 1912, Bohr prepared a draft for Rutherford entitled *On the Constitution of Atoms and Molecules*, which faced directly the problem of atomic stability.

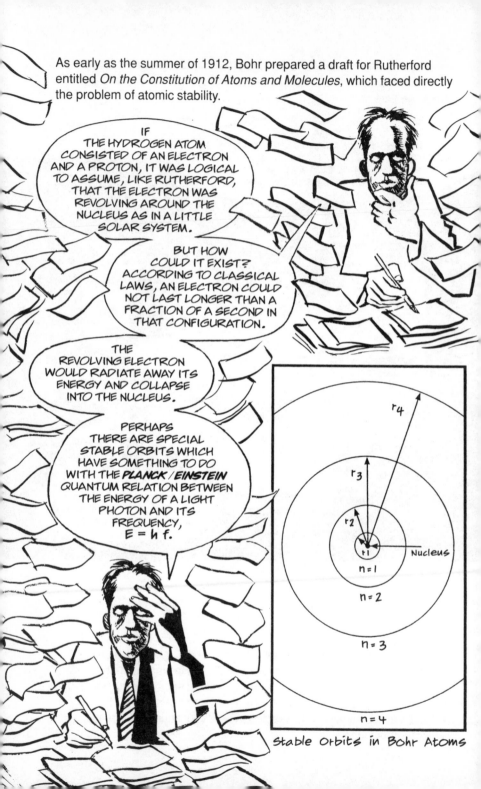

IF THE HYDROGEN ATOM CONSISTED OF AN ELECTRON AND A PROTON, IT WAS LOGICAL TO ASSUME, LIKE RUTHERFORD, THAT THE ELECTRON WAS REVOLVING AROUND THE NUCLEUS AS IN A LITTLE SOLAR SYSTEM.

BUT HOW COULD IT EXIST? ACCORDING TO CLASSICAL LAWS, AN ELECTRON COULD NOT LAST LONGER THAN A FRACTION OF A SECOND IN THAT CONFIGURATION.

THE REVOLVING ELECTRON WOULD RADIATE AWAY ITS ENERGY AND COLLAPSE INTO THE NUCLEUS.

PERHAPS THERE ARE SPECIAL STABLE ORBITS WHICH HAVE SOMETHING TO DO WITH THE *PLANCK / EINSTEIN* QUANTUM RELATION BETWEEN THE ENERGY OF A LIGHT PHOTON AND ITS FREQUENCY, $E = h f$.

r_4

r_3

r_2

r_1 Nucleus

n = 1

n = 2

n = 3

n = 4

Stable Orbits in Bohr Atoms

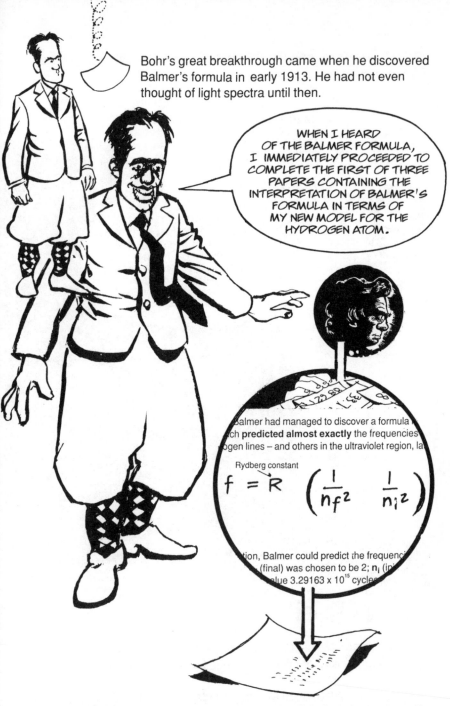

Bohr's great breakthrough came when he discovered Balmer's formula in early 1913. He had not even thought of light spectra until then.

WHEN I HEARD OF THE BALMER FORMULA, I IMMEDIATELY PROCEEDED TO COMPLETE THE FIRST OF THREE PAPERS CONTAINING THE INTERPRETATION OF BALMER'S FORMULA IN TERMS OF MY NEW MODEL FOR THE HYDROGEN ATOM.

Balmer had managed to discover a formula which **predicted almost exactly** the frequencies ...gen lines – and others in the ultraviolet region, la...

Rydberg constant

$$f = R \left(\frac{1}{n_f{}^2} \quad \frac{1}{n_i{}^2} \right)$$

...tion, Balmer could predict the frequenc... (final) was chosen to be 2; n_i (in... ...lue 3.29163×10^{15} cycles...

That event marks the beginning of the quantum theory of atomic structure.

Bohr Meets Nicholson: Quantized Angular Momentum

J.W. Nicholson (1881–1955) had quantized angular momentum, proceeding to calculate the correct value **L = mvR = n (h/2π)** for hydrogen.

Bohr did not seem to need Nicholson's idea to continue his work at that moment. But it proved important, so we should take a careful look at angular momentum.

First: *Linear* Momentum

In our everyday language we use the term *momentum* to refer to something that is difficult to stop once it is moving. In physics, the meaning is the same. In a linear or straight line system with no friction, **a body set in motion will continue in motion unless acted on by an outside force.** This is called the *principle of conservation of momentum* and was known to Galileo even before Newton was born.

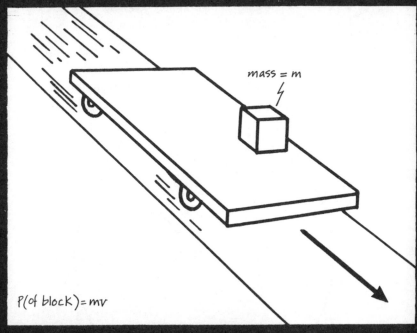

mass = m

P(of block) = mv

THE NUMERICAL VALUE OF THE *LINEAR MOMENTUM* IS DEFINED SIMPLY AS THE PRODUCT OF THE MASS OF THE BODY AND ITS SPEED: p = m v (LINEAR MOMENTUM)

Second: *Angular* Momentum

In a *rotating* system the physics is similar. If a body is set in *rotational* motion in a closed orbit without friction, it will continue undiminished with **constant angular momentum** until acted on by an external torque. The magnitude is simply given by the product of the body's mass, its speed and the radius of the orbit . . .

$$L = m\,v\,r \quad \text{(angular momentum)}$$

where **m** is the mass and **v** the speed around the orbit.

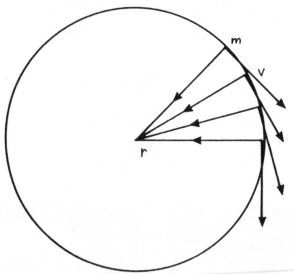

Constant angular momentum (no torque)

In Bohr's model, if an electron is excited from its initial energy state, it can only "jump" to an orbit where its angular momentum will change – increase or decrease – by some whole number times $h/2\pi$.

THIS IS THE CENTRAL PREMISE OF MY SCHEME . . . THE QUANTIZATION OF THE ELECTRON ORBITS IN THE ATOM IN UNITS OF PLANCK'S CONSTANT.

The Bohr Quantum Postulates

Bohr introduced two new postulates to account for the existence of stable electron orbits. In the first, he justified the use of the nuclear atom in defiance of classical objections.

Bohr's First Postulate

AN ATOM *CAN* EXIST IN ANY ONE OF SEVERAL SPECIAL ORBITS WITH NO EMISSION OF RADIATION, CONTRARY TO THE EXPECTATIONS OF CLASSICAL PHYSICS.

THESE ORBITS ARE CALLED *STATIONARY* STATES AND ARE CHARACTERIZED BY VALUES OF ORBITAL ANGULAR MOMENTUM GIVEN BY . . .

$$L = m \vee r = n \, (h/2\pi)$$

This is the quantum orbital condition.

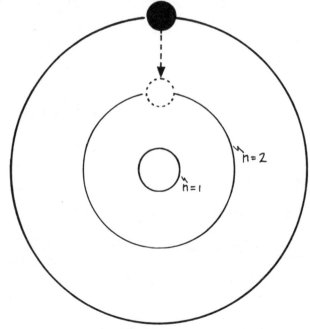

Electron in stationary state ready to make quantum jump

$n = 2$

$n = 1$

The angular momentum **L** cannot take on **any** arbitrary value, as is the usual case in classical physics, but only certain values. **L = 1 (h/2π)** in the first orbit; **L = 2 (h/2π)** in the second orbit; **L = 3 (h/2π)** in the third . . . and so on. Only orbits in which **L** is a whole multiple of the quantized unit **h/2π** are allowed. (This integer **n** is called the *principal quantum number*.)

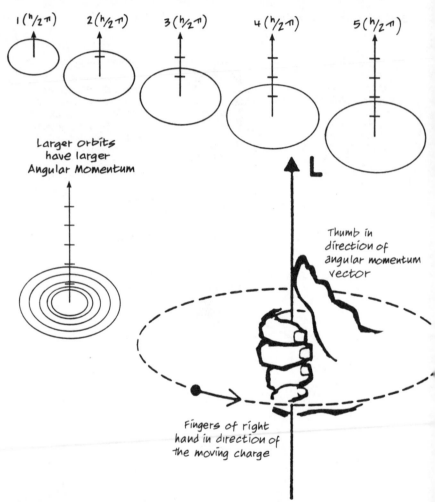

$1\left(^h/_2\pi\right)$ $2\left(^h/_2\pi\right)$ $3\left(^h/_2\pi\right)$ $4\left(^h/_2\pi\right)$ $5\left(^h/_2\pi\right)$

Larger orbits have larger Angular Momentum

L

Thumb in direction of angular momentum vector

Fingers of right hand in direction of the moving charge

What's the fundamental quantum unit, h or h/2π?
First we saw that light can only exist as finely-divided units of energy **E = h f** (frequency). Now we find the angular momentum is also quantized, but this time in units of **h/2π**. So what is the difference? Where does the factor **2π** come from? Why is angular momentum quantized differently from energy? An intriguing question, to be answered soon!

Mixing Classical and Quantum Physics

If the angular momentum of an orbiting body is known – in this case *postulated* – it is a simple matter to compute the radius and the energy of the orbit using classical ideas. Bohr based his derivation on Newton's planetary model of the solar system to obtain a formula for the radii of the electron orbits . . .

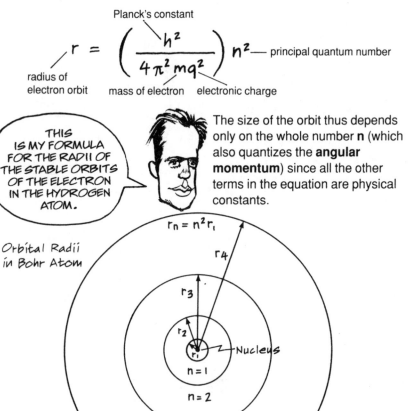

Planck's constant

$$r = \left(\frac{h^2}{4\pi^2 mq^2} \right) n^2 \text{— principal quantum number}$$

radius of electron orbit   mass of electron   electronic charge

THIS IS MY FORMULA FOR THE RADII OF THE STABLE ORBITS OF THE ELECTRON IN THE HYDROGEN ATOM.

The size of the orbit thus depends only on the whole number **n** (which also quantizes the **angular momentum**) since all the other terms in the equation are physical constants.

Orbital Radii in Bohr Atom

$$r_n = n^2 r_1$$

r_4

r_3

r_2

r_1 — Nucleus

n = 1

n = 2

n = 3

n = 4

The smallest radius is for **n** = 1, when its value is 5.3 x 10^{-9}m or 5.3 nanometres. This value is close to modern estimates of the size of the atom based on actual measurements. At this value, called the *Bohr radius*, the energy of the hydrogen atom is a minimum and the atom is in its **ground state**.

Bohr's Second Postulate

Continuing his analogy of the atom as a mini solar system, Bohr could easily calculate the energy of each orbit once the radius was known. He could then use the energy **difference** between stationary states to determine the frequencies of the emission and absorption of light. This led to his *second postulate* . . .

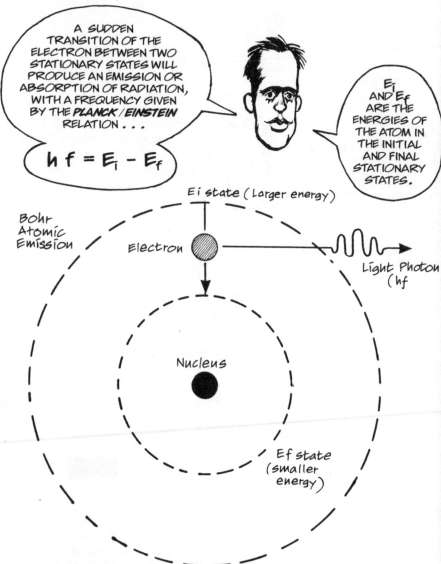

A SUDDEN TRANSITION OF THE ELECTRON BETWEEN TWO STATIONARY STATES WILL PRODUCE AN EMISSION OR ABSORPTION OF RADIATION, WITH A FREQUENCY GIVEN BY THE *PLANCK / EINSTEIN* RELATION . . .

$$h f = E_i - E_f$$

E_i AND E_f ARE THE ENERGIES OF THE ATOM IN THE INITIAL AND FINAL STATIONARY STATES.

E_i state (larger energy)

Bohr Atomic Emission

Electron

Light Photon (hf

Nucleus

E_f state (smaller energy)

This is the quantum transition condition.

Bohr Derives the Balmer Formula

From these postulates, Bohr set out to derive Balmer's formula (already known to give the correct values for the line spectra of hydrogen) using his new atomic model. He mixed classical and quantum physics together to obtain . . .

$$f = \frac{2\pi^2 mq^4}{h^3}\left(\frac{1}{n_f^2} - \frac{1}{n_i^2}\right)$$

This was exactly the same formula Balmer had obtained for the frequencies of hydrogen, if the constant term **R** in Balmer's equation (called the *Rydberg constant*) could be shown to be equal to: $R = (2\pi^2 mq^4/h^3)$.

Using values for **q**, **m** and **h** available in 1914, Bohr calculated: $R = 3.26 \times 10^{15}$ cycles/sec, within a few percent of Balmer's value.

Bohr had **derived the Balmer formula** (which was known to give all the correct hydrogen spectra) from a physical theory based on electrons orbiting the nucleus. A remarkable result.

Bohr could now draw an energy diagram based on physical orbits in the atom to show how the various spectral series originate. Had the young Dane solved the riddle of atomic structure? Would his model work – *i.e.* *predict the spectra* – for all the other elements?

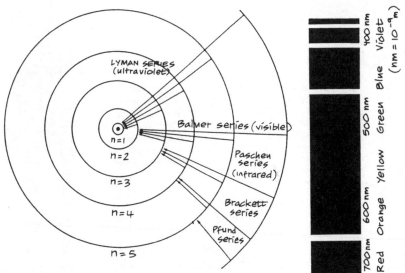

Bohr orbital transmissions for Balmer series and corresponding spectral lines

A Closer Look at Spectra . . . and More Lines

Soon extra spectral lines appeared, even in simple hydrogen, and Bohr's model was being challenged. As more careful measurements of the hydrogen spectra became available, it was obvious that more structure in the atom was necessary. There seemed to be more possible states for the electron than Bohr's simple circular orbits – with only one quantum number – would allow. But a renowned theorist came to the rescue.

Sommerfeld's elliptical orbits

Arnold Sommerfeld (1868–1951), the great theoretician and teacher in Munich.

I EXTENDED BOHR'S IDEAS TO THE CASE OF *ELLIPTICAL ORBITS* AND EXPLAINED THESE EFFECTS.

AFTER ALL, THE MOST GENERAL ASPECT OF ORBITAL MOTION IS *ELLIPTICAL* AND THE CIRCLE IS A SPECIAL CASE.

Johannes Kepler (1571–1630) had done the same to explain the deviations from circularity in the motion of the planet Mars in the light of Tycho Brahe's accurate measurements.

Another Quantum Number Added, k

In spite of the outbreak of the so-called Great War, papers were transmitted secretly from Munich to Copenhagen in which Sommerfeld described elliptical orbits of different shape with the same values of **n**.

THIS RESULTED IN DIFFERENT VALUES OF ENERGY IN THE STATIONARY STATES WITH SLIGHTLY LARGER OR SMALLER ENERGY TRANSITIONS . . . AND THE RESULTING MULTIPLE SPECTRAL LINES.

Again, only certain values of the shapes of the orbit were allowed. Another quantum number was introduced, **k** . . . also quantized in units of **h/2π**.

The Zeeman Effect . . . and Still More Lines

As early as the 1890s, the Dutchman **Pieter Zeeman** (1865–1943) had shown that extraneous spectral lines appeared when the excited atoms were placed in a magnetic field. A true atomic theory would have to explain this phenomenon, which became known as the *Zeeman effect*. Sommerfeld had an answer.

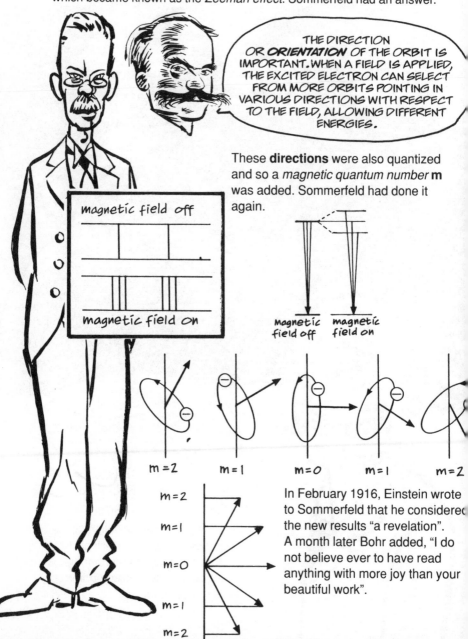

THE DIRECTION OR **ORIENTATION** OF THE ORBIT IS IMPORTANT. WHEN A FIELD IS APPLIED, THE EXCITED ELECTRON CAN SELECT FROM MORE ORBITS POINTING IN VARIOUS DIRECTIONS WITH RESPECT TO THE FIELD, ALLOWING DIFFERENT ENERGIES.

These **directions** were also quantized and so a *magnetic quantum number* **m** was added. Sommerfeld had done it again.

magnetic field off

magnetic field on

magnetic field off   magnetic field on

m = 2   m = 1   m = 0   m = 1   m = 2

m = 2
m = 1
m = 0
m = 1
m = 2

In February 1916, Einstein wrote to Sommerfeld that he considered the new results "a revelation". A month later Bohr added, "I do not believe ever to have read anything with more joy than your beautiful work".

Three Quantum Numbers, n, k, m

With Sommerfeld's calculations to back him up, Bohr worked out a series of selection rules for atomic transitions on the basis of **three** quantum numbers . . . the size of the orbit (**n**), the shape of the orbit (**k**), and the direction in which the orbit is pointing (**m**).

Each separate energy state could now be assigned a distinct set of these integral numbers, **n**, **k**, and **m**, and transitions between these states would produce the observed spectral lines.

OH NO! WILL THIS NEVER CEASE?

Was the Bohr-Sommerfeld scheme now enough to explain **all** the lines observed in the hydrogen spectra? Well no, not quite. Something was still missing. Yet another quantum number was needed to explain fully the magnetic effects.

Wolfgang Pauli: the Anomalous Zeeman Effect, Electron Spin and the Exclusion Principle

The explanation of the magnetic splitting of the spectral lines, reported by Zeeman in 1894, was one of the big successes of the Bohr-Sommerfeld orbits. But later magnetic results produced more lines, and the physicists were stumped. They called this the *Anomalous Zeeman Effect* (AZE).

> BUT IT WASN'T ANOMALOUS AT ALL. THEY JUST COULDN'T UNDERSTAND IT.

In 1924–5, everyone was mystified by the AZE, not the least of whom was the Swiss theoretician **Wolfgang Pauli** (1900–58). In fact, it bothered him so much that it inspired one of many stories people tell about him, most of which are probably true.

Pauli had accepted an invitation to work with Bohr in Copenhagen and wrote two papers on the AZE, neither of which satisfied him. During this stay in 1922 and 1923, he was often depressed and agitated by his lack of progress with this problem. One day, a colleague met Pauli strolling aimlessly in the beautiful streets of Copenhagen . . .

Pauli began his schooling in his native Vienna, where as a teenager he was already advanced in mathematics and physics. In 1918 he enrolled in the University of Munich and with the encouragement of his professor, Sommerfeld, he published a review article on general relativity which became legendary when Einstein wrote: *Whoever studies this mature and grandly conceived work might not believe its author is only 21 years old.*

The Pauli Effect

Pauli did his Ph.D thesis under Sommerfeld in 1921 on the quantum theory of ionized hydrogen. He went for half a year as assistant to Born in Göttingen and then to Hamburg as a *privatdozent*. From that period date the first occurrences of the *Pauli effect* (not to be confused with the *Pauli principle*) . . .

Whenever he entered a laboratory, something would go badly wrong with the experimental apparatus! (Pauli effect.)

It was an accepted fact that theorists were hopeless with experiments. But Pauli was such an *exceptional* theorist that just his presence alone would cause equipment to fall apart. He would relate with hilarity how his friend at Hamburg, the well-respected experimentalist **Otto Stern** (1888–1969), would consult him only through the closed door leading to his laboratory.

The Anomalous Zeeman Effect – which had bothered Pauli so much in Copenhagen – eventually led to Pauli's being immortalized as one of the major contributors to quantum theory.

Pauli's "Hidden Rotation" and the Spinning Electron

Pauli made a hypothesis that a **hidden rotation** produces the extra angular momentum responsible for the AZE. He proposed a **fourth** quantum number with **two** values, just what was needed to explain the perplexing AZE.

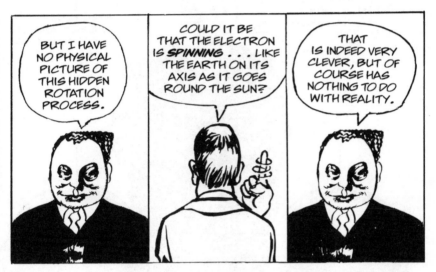

Meanwhile, two young Dutch physicists, **George Uhlenbeck** and **Sam Goudsmit**, had the same idea. Their professor, **Paul Ehrenfest**, was more sympathetic and sent their paper for publication.

It was soon shown that the mysterious results of the AZE were due to the electron's spinning, which gave it extra angular momentum.

There was one troublesome aspect of the spin discovery which needs to be mentioned, since it leads so inevitably to the new quantum theory to follow a year later. The angular momentum of the spinning electron turned out to be only **one-half of the normal value h/2π of atomic orbits**, so-called **spin 1/2**.

This is another example of a semi-classical concept which didn't quite work (e.g. the electron would have to spin around *twice* to get back to its starting point!).

Pauli's Exclusion Principle

The initial puzzle of atomic structure had been why all the electrons do not simply fall into the ground state. To explain why this doesn't happen, Pauli had proposed that each atomic state (a set of three quantum numbers) contained **two** electrons and needed its own exclusive orbit. This was given the fancy title of *space quantization*.

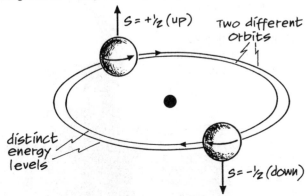

Now with the double-valued spin concept, Pauli was able to make the final pronouncement on his **exclusion principle** . . .

EACH QUANTUM STATE IN THE ATOM IS NOT LIMITED TO **TWO** ELECTRONS BUT ONLY **ONE** ELECTRON. THERE ARE THUS FOUR QUANTUM NUMBERS, COUNTING SPIN UP OR DOWN, FOR EACH DISTINCT ENERGY LEVEL.

IF A STATE IS OCCUPIED, THE NEXT ELECTRON MUST GO TO AN EMPTY HIGHER ENERGY STATE, FILLING UP THE EMPTY STATES FROM THE LOWEST ENERGY TO HIGHER ENERGY. THIS IS WHAT KEEPS THE ATOM FROM ALWAYS COLLAPSING TO ITS LOWEST OR **GROUND STATE** AND GIVES EACH ELEMENT ITS CHARACTERISTIC STRUCTURE.

THE FACT THE ELECTRONS CANNOT ALL GET ON TOP OF EACH OTHER MAKES TABLES AND EVERYTHING ELSE SOLID.

99

Unlike his earlier hypothesis restricted to the outer (or *valence*) electrons to explain the AZE, Pauli now proposed that this principle applied to **all** electrons and **all** atoms. With this simple yet profound principle, the quantum states for any atom could now be constructed and the form of the periodic table of the elements could be understood from first principles.

The Periodic Table: Mendeléev
The periodicity of the elements had been known since the 1890s when the Russian **Dimitri Mendeléev** (1834–1907) invented a visual aid for students struggling with organic chemistry.

I HAD REALIZED THAT THE CHEMICAL PROPERTIES OF THE ELEMENTS WERE REPEATED IF ARRAYED IN A TABLE OF ROWS AND COLUMNS ACCORDING TO INCREASING ATOMIC NUMBER.

ОПЫТЪ СИСТЕМЫ ЭЛЕМЕНТОВЪ.

ОСНОВАННОЙ НА ИХЪ АТОМНОМЪ ВѢСѢ И ХИМИЧЕСКОМЪ СХОДСТВѢ.

			Ti = 50	Zr = 90	? = 180.
			V = 51	Nb = 94	Ta = 182.
			Cr = 52	Mo = 96	W = 186.
			Mn = 55	Rh = 104,4	Pt = 197,4.
			Fe = 56	Ru = 104,4	Ir = 198
		Ni = Co = 59		Pl = 106,6	Os = 199.
H = 1			Cu = 63,4	Ag = 108	Hg = 200
	Be = 9,4	Mg = 24	Zn = 65,2	Cd = 112	
	B = 11	Al = 27,4	? = 68	Ur = 116	Au = 197?
	C = 12	Si = 28	? = 70	Sn = 118	
	N = 14	P = 31	As = 75	Sb = 122	Bi = 210?
	O = 16	S = 32	Se = 79,4	Te = 128?	
	F = 19	Cl = 35,5	Br = 80	I = 127	
Li = 7	Na = 23	Ca = 40	Sr = 87,6	Ba = 137	Pb = 207.
		? = 45	Ce = 92		
		?Er = 56	La = 94		

This periodicity remained a mystery until Pauli's exclusion principle in 1925 gave a truly fundamental explanation. However, Niels Bohr explained it before Pauli's discovery, using his orbital model of the atom.

Bohr's Explanation of the Periodic Table

The periodic table, rather than the explanation of Balmer's spectra, was Bohr's main concern when he began his atomic studies in 1913. He did it with great physical intuition and the details of his orbital model.

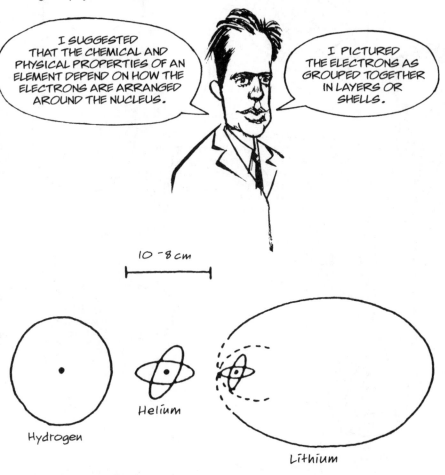

Each shell can contain no more than a certain number of electrons and the chemical properties are related to how nearly full or empty a shell is. For example, full shells are associated with chemical stability. So the electron shells in the inert gases (helium, neon, argon, etc) are assumed to be completely filled.

Bohr began with the observation that the element hydrogen (with 1 electron) and lithium (with 3 electrons) are somewhat alike chemically. Both have valence of one and both enter into compounds of similar types, for example hydrogen chloride, HCl, and lithium chloride, LiCl.

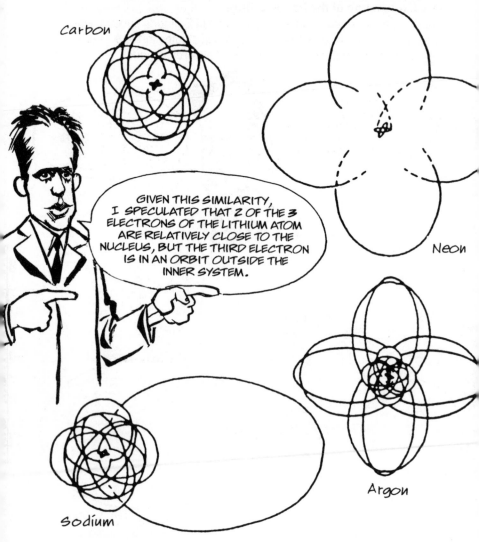

Thus, the lithium atom may be crudely pictured as being like a hydrogen atom. This similar physical structure, then, is the reason for the similar chemical behaviour. So, the first shell has 2 electrons and the third electron goes into the next, or outer, shell.

Closed Shells and Inert Gases

Sodium (with 11 electrons) is the next element in the periodic table that has chemical properties similar to those of hydrogen and lithium. This similarity suggests that the sodium atom also is *hydrogen-like* in having a central core about which one electron revolves. For sodium, then, the eleventh electron must be in an outer shell, so the second shell has 8 electrons.

These qualitative ideas led Bohr to a consistent picture of electrons arranged in groups, or shells, around the nucleus. Hydrogen, lithium, sodium and potassium each have a single electron around a core which is very much like the preceding element, an inert gas. This outlying electron is expected to be easily *involved* with nearby atoms, and this agrees with the facts.

Bohr carried through a complete analysis along these lines and in 1921 proposed the form of the periodic table shown below. Bohr's table is still useful today, an example of a physical theory providing a plausible basis for understanding chemistry.

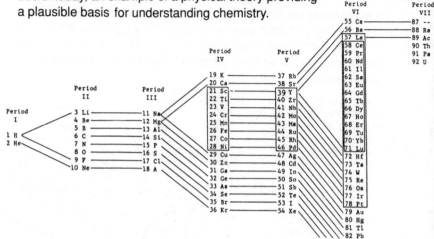

But it was Pauli who gave the fundamental underpinning for Bohr's "physical" periodic chart. His Exclusion Principle (that each and every electron must have its own set of quantum numbers) automatically produces the magic numbers 2, 8, 18, etc., which Bohr devised for his shells. This is the first indication of the fact that each electron in an atom "knows the address" of every other electron and takes its unique space in the atom's structure. (More on this *connectedness* later.)

The table below shows how the Exclusion Principle generates the **magic numbers** (i.e. how many electrons are in each orbit or *shell*). The range of values for quantum numbers **k** and **m** can be inferred from the diagrams on pages 91–3. The 4th quantum number – determined from the AZE – is **s**, the electron spin, which can have only values **up** or **down**. In the table, Bohr's shells correspond to orbits designated by the principle quantum number **n**.

	n	poss k	possible m	poss s	total states
1st shell	1	1	0	± 1/2	2 = 2
	2	1	0	± 1/2	2
2nd shell	2	2	−1, 0, 1	± 1/2	6 = 8
	3	1	0	± 1/2	2
3rd shell	3	2	−1, 0, 1	± 1/2	6 = 18
	3	3	−2, −1, 0, 1, 2	± 1/2	10

The Wave/Particle Duality

Before embarking on a radically new way of viewing electrons in atoms, it is important to understand the properties of waves and to consider the physicist's most perplexing paradox.

Is the fundamental nature of radiation and matter described better by a wave or a particle representation? Or do we need both?

For the origins of the **wave**/**particle** controversy, we must go back to the days when Isaac Newton and the Dutch physicist **Christiaan Huygens** (1629–95) argued about the nature of light.

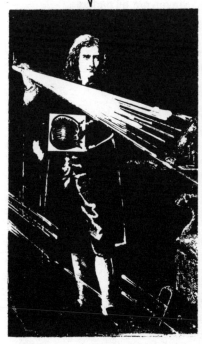

So, who is right? And what are the arguments for a wave theory of light?

Properties of Waves

Think of a pulse transmitted along a stretched, flexible string. This is the simplest kind of wave motion.

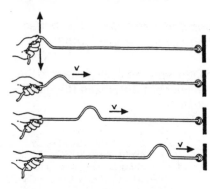

Now consider pulses generated at each end of the string, travelling toward each other. What happens when they overlap demonstrates an important property unique to waves, called **superposition** (which does not occur for particles).

Superposition
If two pulses on a string travel past a particular point at the same time, the total displacement of the string is the sum of the individual displacements.

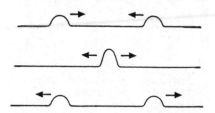

Note that if the pulses have the same size and shape but opposite polarity, they cancel completely at the common point (the energy goes into the motion of the string) and pass right through each other.

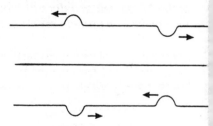

Periodic Waves
Periodic waves occur if one pulse follows another in regular succession. Sound waves, water waves and light waves are all periodic.

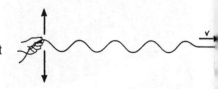

Wave Speed

The speed (**v**), wavelength (**λ**) and frequency (**f**) of a wave are related in a simple way: **v = fλ**. This is obvious from the fact that frequency is the number of waves per second and **λ** is the length of the wave.

Interference: the Double-Slit Experiment

Consider the classic double-slit experiment. If two identical periodic waves arrive at the same point out of phase, i.e. separated by **exactly one-half wavelength**, then **destructive interference** takes place and the waves cancel out (e.g. for light, a dark spot occurs). If the separation is **exactly one whole wavelength**, **constructive interference** takes place and a bright spot appears (i.e. for light).

The double-slit experiment was first reported by **Thomas Young** (1773–1829) in 1801. His demonstration of interference by alternate bright and dark lines was taken to be clear evidence for the wave nature of light. See for yourself in Young's original sketch reproduced here.

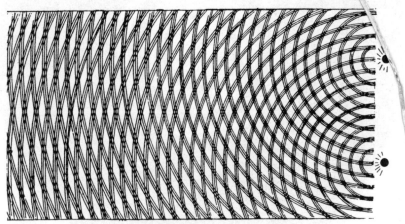

Thomas Young's original drawing showing interference.
The drawing can best be seen by placing your eye near the right edge and sighting at a grazing angle.

Double-slit interference

E — constructive interference produces bright line (waves 'in' phase)

C — Destructive interference produces dark line (waves 'out of' phase)

S — constructive interference produces bright line (waves 'in' phase)

Diffraction and Interference

Diffraction, the bending of waves around an edge, can also cause interference patterns. When a point source of light (or any other kind of wave) passes through a small circular hole of a similar size to the wavelength, diffraction from the edges of the opening spreads the light into a large disk and interference occurs.

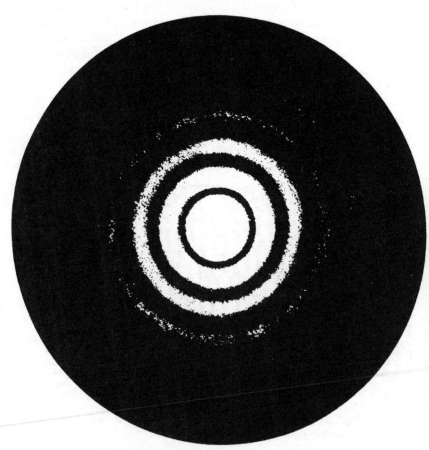

The pattern is shown in the photograph. Although the wave paths are more complicated than the double-slit experiment, the principles are the same. We shall see this pattern again, the unambiguous evidence for waves.

In addition to these interference effects, further evidence for the wave nature of light was demonstrated by Maxwell's electromagnetic wave theory of 1865. The 19th century classical physicists were satisfied. Light consisted of waves.

Einstein . . . a Lone Voice

But as the 20th century unfolded, the young Einstein reintroduced the idea of *corpuscles* to explain the photoelectric effect (see page 44). A few years later, in 1909, he applied his powerful new method of statistical fluctuations to Planck's black-body law and showed that two distinct terms appeared, indicating a *duality* . . .

Einstein was alone in his concern over this problem. No one believed in *photons*. Not for the first time, he was ahead of his contemporaries in dealing with some of the ambiguity of quantum theory, at least for light radiation.

But even he wasn't ready for the shock that came from Paris in 1924. Fortunately, he was contacted immediately. His opinion was urgently needed! 109

A French Prince Discovers Matter Waves

In 1923, a graduate student at the Sorbonne in Paris, Prince Louis de Broglie (1892–1987) introduced the astounding idea that particles may exhibit wave properties. De Broglie had been greatly influenced by Einstein's arguments that a duality may be necessary in understanding light.

In his doctoral thesis of 1924, de Broglie wrote . . .

IT WOULD SEEM THAT THE BASIC IDEA OF THE QUANTUM THEORY IS THE IMPOSSIBILITY OF IMAGINING AN ISOLATED QUANTITY OF ENERGY WITHOUT ASSOCIATING IT WITH A CERTAIN FREQUENCY . . .

HOWEVER, IT IS DIFFICULT TO UNDERSTAND PRECISELY THE PHYSICAL SENSE OF THE FREQUENCY IN THE EINSTEIN EQUATION . . .

$$E = (h)(f)$$

ENERGY EQUALS PLANCK'S CONSTANT TIMES FREQUENCY

BUT IT APPARENTLY DESCRIBES A CERTAIN INTERNAL "CYCLIC PROCESS".

He was deeply impressed by Einstein's particles of light which could cause the photoelectric effect (knock electrons out of a metal) while managing to carry this "periodic" information to produce interference effects in a different context, like the double-slit experiment.

Then came the blockbuster. In the first part of his thesis, de Broglie proposed one of the great unifying principles in all of physics . . .

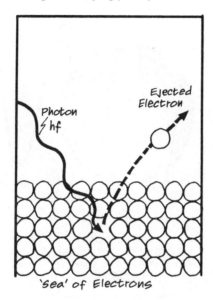

Photon
hf

Ejected
Electron

'Sea' of Electrons

A Photon with measurable frequency (wavelength) interacts with a single Electron

An Associated Wave

What de Broglie did was to assign a frequency, not directly to the internal periodic behaviour of the particle (as he imagined the Einstein photon), but to **a wave which accompanied the particle through space and time**, in such a way that it was always in phase with the "internal" process.

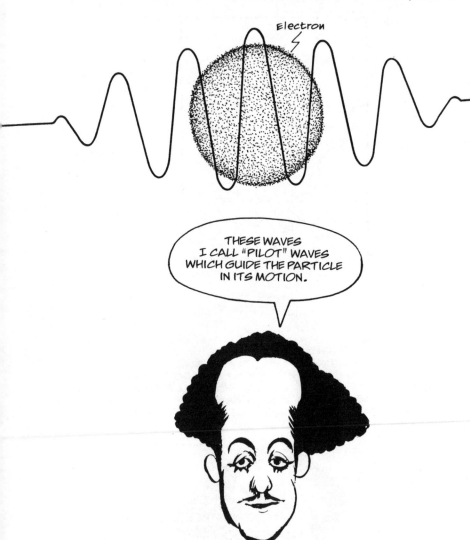

Electron

THESE WAVES I CALL "PILOT" WAVES WHICH GUIDE THE PARTICLE IN ITS MOTION.

Could such waves ever be detected? That is, could these mysterious waves possibly relate to the **actual motion of the particle** and be measured?

Yes, said de Broglie, these waves are not just abstractions. The physically important result of the new radical ideas is that there are **two** velocities associated with the *pilot* waves.

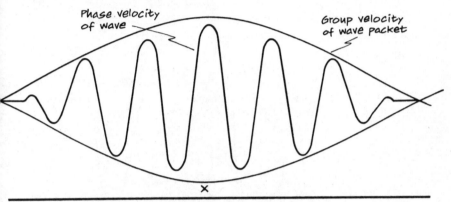

Phase velocity of wave

Group velocity of wave packet

X

A wave packet spread over a distance X

ONE IS THE **PHASE** VELOCITY — THE SPEED AT WHICH A WAVE CREST MOVES — AND THE SECOND IS A **GROUP** VELOCITY — THE SPEED OF THE REINFORCEMENT REGIONS FORMED WHEN MANY WAVES ARE SUPERIMPOSED.

De Broglie identified the group velocity with the velocity of a particle and showed that the reinforcement region displays all the mechanical properties – such as energy and momentum – normally associated with a particle. (This is similar to the way a **pulse** is produced by a superposition of many waves of different frequencies.)

There were more dramatic conclusions to come when he wrote down the simple mathematical relationships describing these ideas which were based on an analogy with photons.

He started with Einstein's famous $E = mc^2$ equation for the total energy content of anything. In this case, photons . . .

$$E = mc^2 = (mc)(c)$$

Now watch de Broglie's series of substitutions . . .

Since mc is just mass times speed, the *momentum*, p, of a photon . . .

$$E = (p)(c) = (p)(f\lambda)$$

using c (speed) = f (frequency) times λ (wavelength) for waves.

Equating $E = h\,f$ from the Planck/Einstein relation to the expression above, we obtain:

$$(h)(f) = (p)(f\lambda)$$

and some simple algebra gives . . . $h/p = \lambda$ (photons).

THIS MEANS THAT IF THE WAVELENGTH OF LIGHT IS *DECREASED*, THE MOMENTUM OF THE INDIVIDUAL LIGHT PHOTONS IS *INCREASED*.

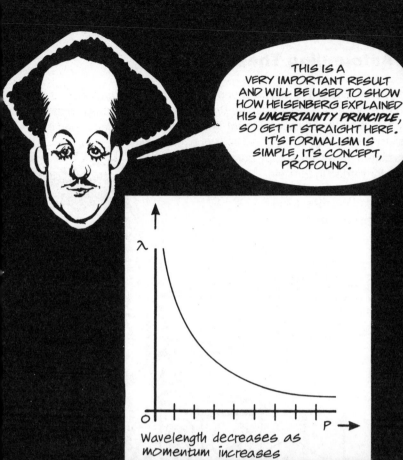

Wavelength decreases as momentum increases

By direct analogy, de Broglie proposed that his relation was true not just for **photons**, but for **electrons** . . . and **all** particles.

$$\lambda = h/p \ldots$$

or (wavelength) = (Planck's constant divided by momentum)

For electrons, the

$$\text{MOMENTUM } p = (m)(v) = \textit{(MASS)(VELOCITY)}$$

can easily be determined in an experimental situation and the *wavelength* could thus be predicted from de Broglie's equation.

To most physicists the concept seemed preposterous. The electron was a **PARTICLE**, known by classical physicists since J.J. Thomson's discovery in 1897!

An Astounding Thesis

These ideas astounded and confounded the examining committee at the University of Paris in 1924 when de Broglie presented his thesis entitled *Researches on the Quantum Theory*. The committee included the eminent physicist **Paul Langevin** (1872–1946) who fortunately had secured an advance copy from de Broglie which he had forwarded to Einstein.

Einstein read the thesis and informed Henrik Lorentz . . .

I BELIEVE DE BROGLIE'S HYPOTHESIS IS THE FIRST FEEBLE RAY OF LIGHT ON THIS WORST OF OUR PHYSICS ENIGMAS.

To the examining committee, he made a profound comment.

. . . DE BROGLIE HAS LIFTED THE GREAT VEIL.

The committee passed him for the Ph.D.

Confirmation of Matter Waves

In just a few years, all of de Broglie's predictions were confirmed by experiment. Remarkably, in defending his thesis against one sceptical member of the examining committee, de Broglie had actually suggested . . .

MATTER WAVES MIGHT BE OBSERVABLE IN CRYSTAL DIFFRACTION EXPERIMENTS LIKE THOSE CARRIED OUT WITH X-RAYS.

In a strange ironic twist, such diffraction patterns were first demonstrated by **G.P. Thomson** (1892–1975) – proving the **wave** property of electrons.

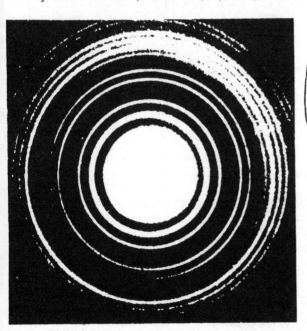

THIS WAS ABOUT 30 YEARS AFTER MY FATHER *J.J. THOMSON* FIRST DEMONSTRATED THE *PARTICLE* PROPERTY OF ELECTRONS.

De Broglie had yet another interesting idea about electron waves in atoms . . . as we can see next.

117

Electron Waves in Atoms

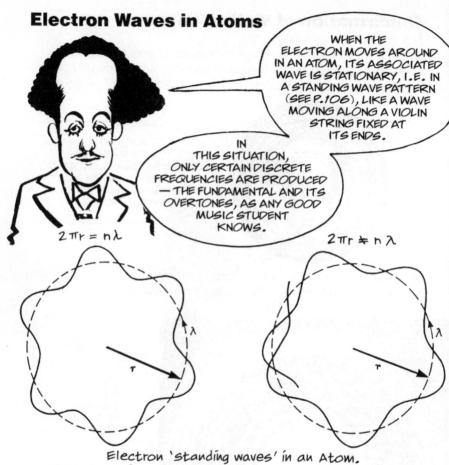

WHEN THE ELECTRON MOVES AROUND IN AN ATOM, ITS ASSOCIATED WAVE IS STATIONARY, I.E. IN A STANDING WAVE PATTERN (SEE P.106), LIKE A WAVE MOVING ALONG A VIOLIN STRING FIXED AT ITS ENDS.

IN THIS SITUATION, ONLY CERTAIN DISCRETE FREQUENCIES ARE PRODUCED — THE FUNDAMENTAL AND ITS OVERTONES, AS ANY GOOD MUSIC STUDENT KNOWS.

$2\pi r = n\lambda$

$2\pi r \neq n\lambda$

Electron 'standing waves' in an Atom.
Only certain wavelengths will fit around a circle.

This is just what Bohr needed in 1913 for his hydrogen atom postulate. (Remember the unexplained 2π factor?) By just fitting a whole number of electron waves along the circumference of the atom, and using de Broglie's relations, Bohr could have given a complete theoretical justification for the orbital quantization. Watch, a little algebra . . .

$$n\lambda = 2\pi r \quad \text{(standing waves)}$$

$$n(h/mv) = 2\pi r \quad \text{(using de Broglie equation)}$$

$$n(h/2\pi) = mvr \quad \text{(quantum orbital postulate)}$$

*Bohr's quantum condition is no longer a **postulate**, it's a reality . . .*

Visualizing the Atom: the "Old Quantum Theory"

The "old quantum theory", resulting in Bohr's orbital model of the atom and its modifications by Sommerfeld, could point to certain real successes: the **hydrogen spectrum**, i.e. the derivation of the Balmer formula; **quantum numbers** and **selection rules** for energy states in an atom; **explanation of the periodic table** of the elements; and the Pauli **Exclusion Principle**.

BUT HOW SHOULD WE NOW THINK OF THE ELECTRON IN THE HYDROGEN ATOM . . . AS A TINY CHARGED PARTICLE CIRCLING THE NUCLEUS, JUMPING FROM ONE ALLOWED ORBIT TO ANOTHER?

OR AS A WAVE ADJUSTING ITS LENGTH JUST ENOUGH TO FIT EXACTLY INTO ONE OF THE ORBITS, SETTING UP A STANDING WAVE PATTERN WITH ITS ELECTRIC CHARGE SOMEHOW DISTRIBUTED AROUND THE CIRCUMFERENCE?

For the moment it doesn't matter. We will need both to continue. But with this ambiguous picture of wave and particle for the electron inside the atom, we are getting closer to the real underlying essence of quantum theory.

Triple Birth of the New Quantum Theory

Now a remarkable report on the end of 25 years of confusion.
During the twelve month period from June 1925 to June 1926, not one, not two, but **three** distinct and independent developments of a complete quantum theory were published . . .
and then shown to be **equivalent**.

THE FIRST —
MATRIX MECHANICS —
BY WERNER HEISENBERG.

THE SECOND —
WAVE MECHANICS —
BY ERWIN SCHRÖDINGER.

THE THIRD —
QUANTUM ALGEBRA —
BY PAUL DIRAC.

The following pages will outline how these discoveries were made and the context which made them possible.
The story begins with Bohr and his new protégé, Werner Heisenberg.

Heisenberg, Genius and Mountain-Climber

Heisenberg (1901–76) grew up in Munich, where his father was professor of Greek at the local university. Always interested in mountain walking, Heisenberg was fortunate that Munich is set at the foot of the Bavarian Alps. He was a brilliant student and an excellent pianist. At secondary school, he had already immersed himself in independent studies of physics.

In the autumn of 1920, immediately after he had enrolled in the University of Munich to study physics with Sommerfeld, he met Wolfgang Pauli. This was the beginning of a lifelong friendship.

Pauli and Heisenberg were both at Göttingen in June 1922 when Heisenberg first met Bohr. Only 20 years old and still working toward his Ph.D, Heisenberg rose to make an objection after one of Bohr's lectures, to which Bohr replied somewhat hesitantly . . .

But Heisenberg had a surprise for Niels Bohr. He hated the imaginary electron orbits in Bohr's atomic model . . .

THEY COULD NEVER BE OBSERVED. WHAT GOOD IS IT TO SPEAK OF *INVISIBLE* ELECTRON PATHS INSIDE *INVISIBLE* TINY ATOMS?

IF AN ATOM CAN'T BE SEEN, THEN IT IS NOT A MEANINGFUL CONCEPT.

In the spring of 1925, he left Copenhagen and returned to Göttingen where Max Born (1882–1970) had made him a *privatdozent* at the age of only 22! In Germany he was bothered by two major irritants: the pollen in the air and the problem of the atomic orbits.

I HAD THIS VERY BAD ATTACK OF HAY FEVER. I COULDN'T EVEN SEE.

I WAS IN A TERRIBLE STATE AND DECIDED TO SEEK BETTER, I.E. POLLEN-FREE, AIR. I LEFT FOR THE ISLAND OF *HELGOLAND* IN THE NORTH SEA.

I WAS EXTREMELY TIRED WHEN I ARRIVED AND MY WHOLE FACE WAS SWOLLEN. THE LANDLADY AT AN INN ASKED IF I HAD BEEN BEATEN BY SOMEBODY.

Heisenberg's Picture of the Atom

Heisenberg hardly slept, dividing his time between inventing quantum mechanics, climbing rocks and memorizing poems by Goethe. He was attempting to work out a **code** for connecting the quantum numbers and energy states in an atom with the experimentally determined frequencies and intensities (brightness) of the light spectra.

MY STARTING POINT WAS TO TREAT THE ATOM, NOT LIKE A LITTLE SOLAR SYSTEM, BUT LIKE A SIMPLE VIRTUAL OSCILLATOR WHICH COULD PRODUCE ALL THE FREQUENCIES OF THE SPECTRUM.

This was similar to what Planck had done on black-body radiation in 1900.

Using the concept which Bohr had called the *correspondence principle*, (where quantum and classical regions overlap), Heisenberg imagined the Bohr atom at very large orbits. There the orbital frequency would equal the radiation frequency and the atom would be like a simple linear oscillator.

He knew how to analyse this problem from classical physics. Familiar quantities like the linear momentum (**p**) and the displacement from equilibrium (**q**) could now be used. Classically, he could solve the equation of motion, then calculate the energy of the particle in the state **n**, the quantized values, E_n.

From the largest orbit – where he could get answers – he then tried to extrapolate *inside* the atom. Here his intuition, some would call it genius, led him to a formula for including all the possible states. **He had broken the spectral code.**

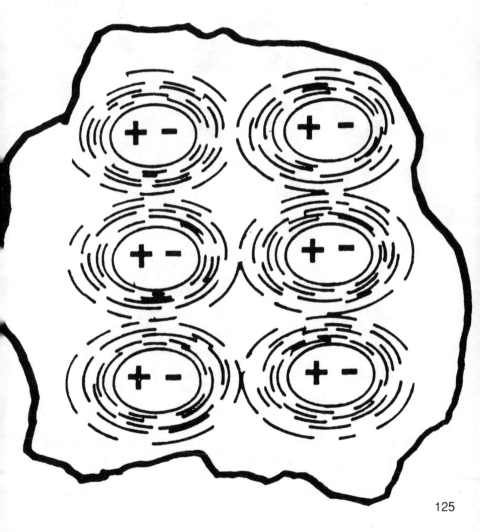

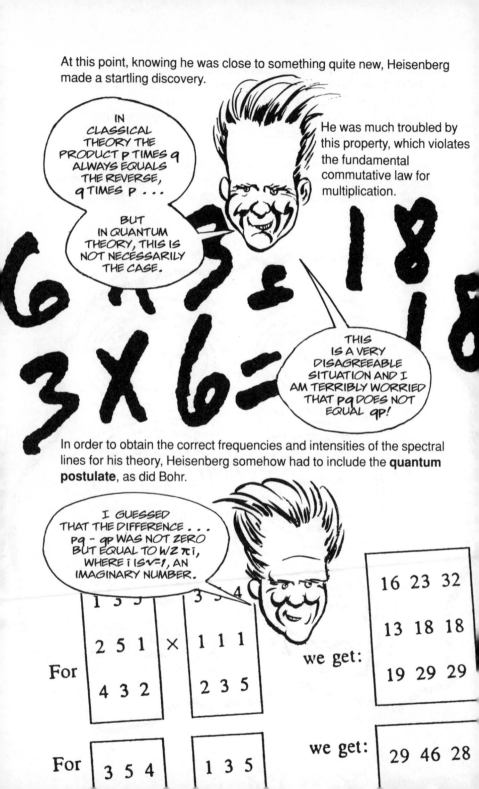

At this point, knowing he was close to something quite new, Heisenberg made a startling discovery.

He was much troubled by this property, which violates the fundamental commutative law for multiplication.

In order to obtain the correct frequencies and intensities of the spectral lines for his theory, Heisenberg somehow had to include the **quantum postulate**, as did Bohr.

That very night on Helgoland, he was able to show that the energy states were **quantized** and **time independent**, i.e. they were *stationary* as in the Bohr atom. He later called this . . .

. . . A GIFT FROM HEAVEN.

IT WAS ABOUT THREE O'CLOCK AT NIGHT WHEN THE FINAL RESULT OF THE CALCULATION LAY BEFORE ME. AT FIRST I WAS DEEPLY SHAKEN, SO EXCITED THAT I COULD NOT THINK OF SLEEP.

SO I LEFT THE HOUSE AND AWAITED THE SUNRISE ON TOP OF A ROCK. THAT WAS "THE NIGHT OF HELGOLAND".

On 19 June, Heisenberg returned to Göttingen and sent his results to Pauli, the invaluable critic. If his theory was correct, he had taken a first step towards killing the concept of orbits. He was now almost fully recovered from *both* his illnesses . . . the hay fever **and** the electron orbits!   127

Max Born and Matrix Mechanics

Pauli's reaction was favourable. So before setting off for a visit to the Cavendish Laboratory in Cambridge and a walking holiday, Heisenberg set the paper before Max Born.

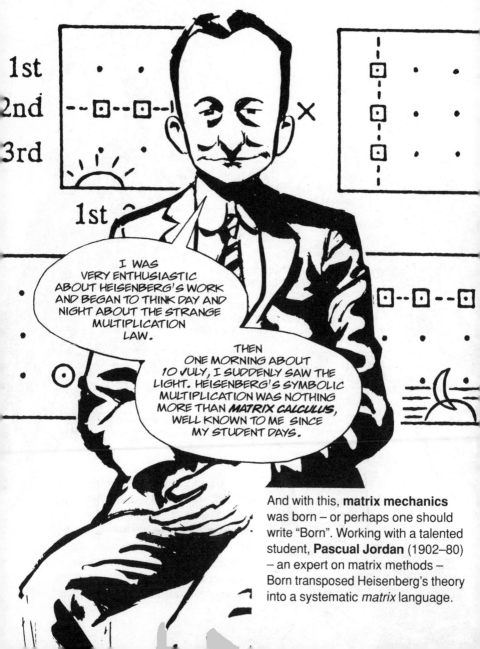

And with this, **matrix mechanics** was born – or perhaps one should write "Born". Working with a talented student, **Pascual Jordan** (1902–80) – an expert on matrix methods – Born transposed Heisenberg's theory into a systematic *matrix* language.

Now the frequencies of the optical spectrum could be represented by an infinite matrix which looks like this . . .

$$f_{m,n} \quad \begin{array}{llllll} f_{11} & f_{12} & f_{13} & f_{14} & f_{15} & f_{16} \quad \text{etc.} \\ f_{21} & f_{22} & f_{23} & f_{24} & f_{25} & f_{26} \quad \text{etc.} \\ f_{31} & f_{32} & f_{33} & f_{34} & f_{35} & f_{36} \quad \text{etc.} \\ f_{41} & f_{42} & f_{43} & f_{44} & f_{45} & f_{46} \quad \text{etc.} \\ \text{etc.} & \text{etc.} & \text{etc.} & \text{etc.} & \text{etc.} & \text{etc.} \quad \text{etc.} \end{array}$$

Since Heisenberg's idea was that the individual oscillators with momentum **p(t)** and displacement **q(t)** vibrate with these frequencies, they will also be infinite matrices.

$$\bullet = \begin{array}{lllll} p_{11} & p_{12} & p_{13} & p_{14} & \text{etc.} \\ p_{21} & p_{22} & p_{23} & p_{24} & \text{etc.} \\ p_{31} & p_{32} & p_{33} & p_{34} & \text{etc.} \\ \text{etc.} & \text{etc.} & \text{etc.} & \text{etc.} & \text{etc.} \end{array} \quad \text{and} \quad \mathbf{q} \begin{array}{lllll} q_{11} & q_{12} & q_{13} & q_{14} & \text{etc.} \\ q_{21} & q_{22} & q_{23} & q_{24} & \text{etc.} \\ q_{31} & q_{32} & q_{33} & q_{34} & \text{etc.} \\ \text{etc.} & \text{etc.} & \text{etc.} & \text{etc.} & \text{etc.} \end{array}$$

Heisenberg's quantum postulate was introduced to obtain the correct frequencies and intensities, each represented by a set of two numbers in *matrix* form.

pq − qp = (h/2πi) I (quantum condition)

I is the unit matrix which looks like this . . .

$$\mathbf{1} = \begin{array}{llll} 1 & 0 & 0 & \text{etc.} \\ 0 & 1 & 0 & \text{etc.} \\ 0 & 0 & 1 & \text{etc.} \\ \text{etc.} & \text{etc.} & \text{etc.} & \text{etc.} \end{array}$$

129

Pauli Shows Matrix Mechanics is Correct

When this condition was added to the classical equation of mechanics **written in matrix form**, a system of equations was obtained which could produce values of the frequencies and relative intensities of spectral lines of atoms. However, . . .

THOUGH I CAN DERIVE ALL THE OLD NEWTONIAN RESULTS WITH MY NEW THEORY, I CAN'T EVEN CALCULATE THE HYDROGEN SPECTRUM.

DON'T WORRY WERNER, I HAVE MASTERED THE COMPLEXITIES OF YOUR NEW MECHANICS ALREADY AND DEDUCED NOT ONLY THE SPECTRUM OF HYDROGEN BUT THE ADDITIONAL LINES PRODUCED BY ELECTRIC AND MAGNETIC FIELDS AS WELL.

Heisenberg had discovered the first complete version of **quantum mechanics**.

But something was different. The new theory came with no visual aids, no model to picture in one's mind. Gone were the intricate electron orbits which Bohr and Sommerfeld had concocted to explain the hydrogen spectra. This was a purely mathematical formalism, difficult to use and impossible to visualize. It simply gave the right answers.

Heisenberg had abandoned all attempts to picture the atom as composed either of particles or waves. He decided that any attempt to draw an analogy between atomic structure and the structure of the classical world was destined to failure.

INSTEAD, I DESCRIBED THE ENERGY LEVELS OF ATOMS PURELY IN TERMS OF *NUMBERS*. SINCE THE MATHEMATICAL DEVICE USED TO MANIPULATE THESE NUMBERS WAS CALLED A MATRIX, MY THEORY WAS CALLED *MATRIX MECHANICS*, A TERM I LOATHED BECAUSE IT WAS SO ABSTRACT.

Later, the spectral patterns of other atoms were also derived. Yet no one knew the **physical significance** of the strange non-commutability, a fundamental part of the theory.

Could it mean that the order in which measurements were made might be important? Could the act of measurement be that critical?

Erwin Schrödinger – Genius and Lover

Meanwhile, other physicists had not given up on the idea of visualizing **all aspects** of the physical universe, of which atomic structure should be a part! Consequently, they did not take well to Heisenberg's matrix mechanics.

In particular, the talented Erwin Schrödinger in Zürich despised the new theory, devoid of pictures and full of mathematical complications.

HERWIG

> I SET OUT TO DEVELOP ANOTHER VERSION BASED ON DE BROGLIE'S CONCEPT OF MATTER WAVES.

> I BELIEVE MY APPROACH IS MORE ACCEPTABLE TO PHYSICISTS AND MARKS A RETURN TO THE CONTINUOUS, VISUALIZABLE WORLD OF CLASSICAL PHYSICS.

Schrödinger was right about the first part, but dead wrong about the second!

Where Werner Heisenberg needed the solitude of mountain walks in the pollen-free air, and Paul Dirac the monastic tranquillity of his college rooms at St. John's, Cambridge, Erwin Schrödinger needed something quite different for **his** inspiration.

Schrödinger was a notorious womanizer, often inspired in his physics work by his most recent love interest. During the Christmas holidays of 1925, he made the most important discovery of his career during a passionate tryst in his favourite romantic hotel in the Austrian Tyrol. He had been thinking about waves.

Schrödinger's Equation

The solution to Schrödinger's equation was a *wave* that described in some magical way the quantum aspects of the system. The physical interpretation of this wave was to become one of the great philosophical problems of quantum mechanics.

I HAVE FOUND AN EQUATION WHICH CAN BE APPLIED TO ANY PHYSICAL SYSTEM IN WHICH THE MATHEMATICAL FORM OF THE ENERGY IS KNOWN.

second derivative with respect to X

energy

potential energy

$$\frac{\partial^2 \psi}{\partial x^2} + \frac{8\pi^2 m}{h^2}\left(E - V\right)\psi = 0$$

position

Schrödinger wave function

The wave itself is represented by the Greek symbol ψ, which to every physicist today means only one thing . . . the solution to Schrödinger's equation. He had taken de Broglie's idea of the wave description of matter very seriously indeed.

Fourier Wave Analysis of Periodic Functions

Although this heading sounds very technical, it is important to say just a few words about Fourier analysis in order to appreciate the delight of the physicists when Schrödinger's equation appeared in January 1926.

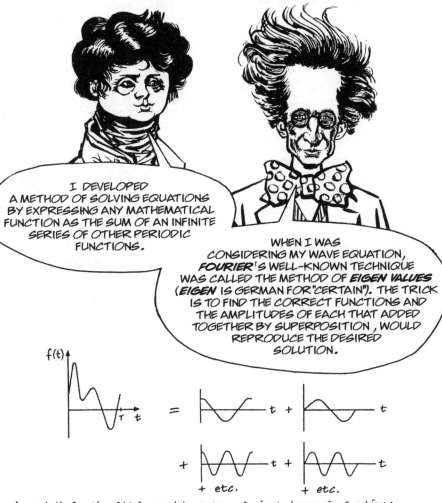

I DEVELOPED A METHOD OF SOLVING EQUATIONS BY EXPRESSING ANY MATHEMATICAL FUNCTION AS THE SUM OF AN INFINITE SERIES OF OTHER PERIODIC FUNCTIONS.

WHEN I WAS CONSIDERING MY WAVE EQUATION, *FOURIER'S* WELL-KNOWN TECHNIQUE WAS CALLED THE METHOD OF *EIGEN VALUES* (*EIGEN* IS GERMAN FOR "CERTAIN"). THE TRICK IS TO FIND THE CORRECT FUNCTIONS AND THE AMPLITUDES OF EACH THAT ADDED TOGETHER BY SUPERPOSITION, WOULD REPRODUCE THE DESIRED SOLUTION.

Any periodic function $f(t)$ is equal to a sum of simple harmonic functions

Thus, the solution of Schrödinger's equation – **the wave function for the system** – was replaced by an infinite series – **the wave functions of the individual states** – which are natural harmonics of each other. That is to say, their frequencies are related in the ratio of whole numbers, or integers. 135

The method is shown by the graphs below. The bold curve indicates the initial function which is then replaced by the sum of the infinite series of the harmonic periodic waves.

Schrödinger's remarkable discovery was that the **replacement waves** described the **individual states of the quantum system** and their **amplitudes** gave the relative importance of that particular state to the whole system.

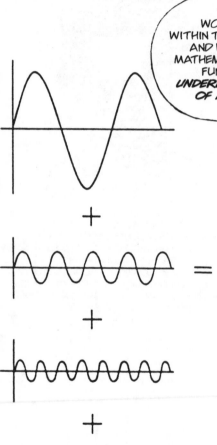

IN OTHER WORDS, CONTAINED WITHIN THE WELL-ESTABLISHED AND WELL-UNDERSTOOD MATHEMATICS OF EIGEN VALUE FUNCTIONS WAS THE *UNDERLYING QUANTIZATION OF ATOMIC SYSTEMS*.

Schrödinger's equation has been universally recognized as one of the greatest achievements of 20th century thought, containing much of physics and, in principle, all of chemistry. It was immediately accepted as a mathematical tool of unprecedented power for dealing with problems of the atomic structure of matter.

Not surprisingly, the work became known as *wave mechanics*.

Visualizing Schrödinger's Atom

What Schrödinger did was reduce the problem of the energy states in an atom to a problem of finding the natural overtones of its vibrating system using Fourier analysis.

The natural frequencies and the number of nodes of *one-dimensional standing waves* (e.g. a violin string) are easy to visualize. This picture can be extended to a *two-dimensional system*, such as the vibrations of a struck drum head. Computer simulation of different vibrational states in a drum gives some indication of what Schrödinger had in mind.

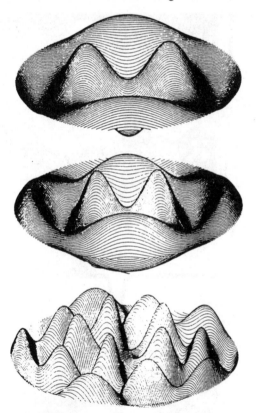

Though it is very difficult to visualize *three-dimensional vibrating systems* in something like the hydrogen atom, the one-dimensional and two-dimensional pictures should be helpful.

The integers called *quantum numbers* by Bohr, Sommerfeld and Heisenberg were now related in a natural way to **the number of nodes in a vibrating system**. 137

The Balmer Formula, the Zeeman Effect and All That

Soon it was shown that Schrödinger's theory gave a complete description of the spectral lines in the hydrogen atom, reproducing again the *touchstone* Balmer formula. In addition, the splitting in electric and magnetic fields also popped right out of the wave equation.

Schrödinger was thus able to observe that the integers (number of nodes) derived from a three-dimensional wave solution precisely correspond to the three quantum numbers **n**, **k** and **m** from the old quantum theory.

Schrödinger: a Return to Classical Physics?

In spite of the innovation of his breakthrough in quantum theory, the Austrian mathematical physicist was from the traditional school of physics. He loathed the concept of discontinuous quantum jumps within the atom proposed by Bohr. Now he had a mathematical system which could explain the spectral lines without the need to postulate these despicable quantum jumps. He made an analogy to sound waves . . .

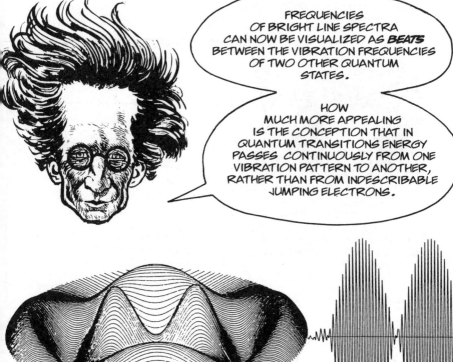

> FREQUENCIES OF BRIGHT LINE SPECTRA CAN NOW BE VISUALIZED AS *BEATS* BETWEEN THE VIBRATION FREQUENCIES OF TWO OTHER QUANTUM STATES.

> HOW MUCH MORE APPEALING IS THE CONCEPTION THAT IN QUANTUM TRANSITIONS ENERGY PASSES CONTINUOUSLY FROM ONE VIBRATION PATTERN TO ANOTHER, RATHER THAN FROM INDESCRIBABLE JUMPING ELECTRONS.

Schrödinger intended to use his new discovery as a pathway **back** to a physics based on continuum processes undisturbed by sudden transitions. He was proposing an essentially classical *theory of matter waves* that would have the same relationship to mechanics that Maxwell's *theory of electromagnetic waves* had to optics.

Who Needs Particles Anyway?

Schrödinger even began to doubt the **existence** of particles.

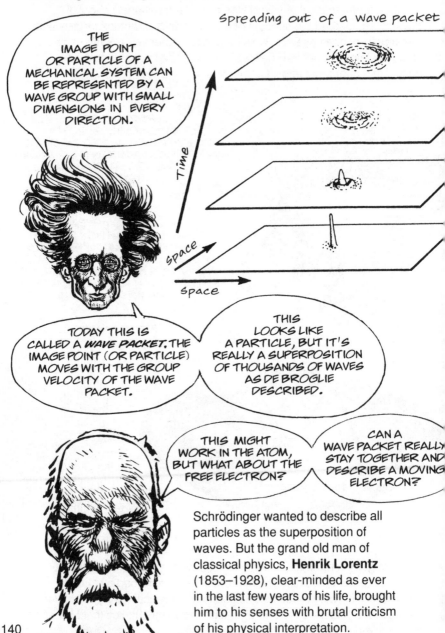

Spreading out of a wave packet

THE IMAGE POINT OR PARTICLE OF A MECHANICAL SYSTEM CAN BE REPRESENTED BY A WAVE GROUP WITH SMALL DIMENSIONS IN EVERY DIRECTION.

Time

Space

Space

TODAY THIS IS CALLED A *WAVE PACKET.* THE IMAGE POINT (OR PARTICLE) MOVES WITH THE GROUP VELOCITY OF THE WAVE PACKET.

THIS LOOKS LIKE A PARTICLE, BUT IT'S REALLY A SUPERPOSITION OF THOUSANDS OF WAVES AS DE BROGLIE DESCRIBED.

THIS MIGHT WORK IN THE ATOM, BUT WHAT ABOUT THE FREE ELECTRON?

CAN A WAVE PACKET REALLY STAY TOGETHER AND DESCRIBE A MOVING ELECTRON?

Schrödinger wanted to describe all particles as the superposition of waves. But the grand old man of classical physics, **Henrik Lorentz** (1853–1928), clear-minded as ever in the last few years of his life, brought him to his senses with brutal criticism of his physical interpretation.

It was soon shown that the wave function **does** spread out as time increases. Clearly, Schrödinger was wrong and Lorentz was right!

So **what is** the relationship between the particle's wave function and the particle itself? Tough question. It was the final issue to be resolved in the development of wave mechanics.

Two Theories, One Explanation

Schrödinger wondered if there was any relationship between his own theory and Heisenberg's matrix mechanics. At first he could see no connection. But in the last week of February 1926, he found a remarkable result of his own analysis.

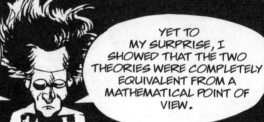

I AM REPELLED BY THE HEISENBERG FORMALISM BECAUSE OF THE DIFFICULT ALGEBRA INVOLVED AND THE LACK OF AN *ANSCHAULICHKEIT*, A VIEWPOINT OR PICTURE.

YET TO MY SURPRISE, I SHOWED THAT THE TWO THEORIES WERE COMPLETELY EQUIVALENT FROM A MATHEMATICAL POINT OF VIEW.

One was based on a clear conceptual wave model of atomic structure and the other claimed that such a model was meaningless. Yet both gave the same results. Very strange indeed!

Schrödinger's equation was here to stay. In 1987 the equation appeared in its final form on the first day postmark of the Austrian stamp commemorating Schrödinger's 100th anniversary.

Schrödinger Meets Heisenberg

Ir July 1926 Schrödinger lectured in Munich at Sommerfeld's weekly colloquium. Heisenberg was in the audience.

Schrödinger finished speaking, his equation on the blackboard. *Are there any questions?* . . .

Max Born: the Probability Interpretation of ψ

Schrödinger had decided that **ψ** represented a "shadow wave" that somehow indicated the position of the electron. Then he changed his mind, saying it was the "density of the electronic charge". Truthfully, he was confused.

A more acceptable idea was developed by Max Born in the summer of 1926. He wrote a paper on collision phenomena, in which he introduced the **quantum mechanical probability**.

> **Ψ** IS THE **PROBABILITY AMPLITUDE** FOR AN ELECTRON IN THE STATE **n** TO SCATTER INTO THE DIRECTION **m**. IT IS, IN A SENSE, ITS OWN INTENSITY WAVE.

> WHEN IT IS SQUARED AND THE ABSOLUTE VALUE IS TAKEN, IT TURNS OUT TO BE A **PHYSICAL PROBABILITY** OF THE ASSOCIATED PARTICLE'S PRESENCE.

Electron Ψn
Electron in state 'n'

Ψm
Electron scattered into state '

One month later, Born stated that the probability of the **existence of a state** is given by the **square** of the normalized amplitude of the individual wave function (i.e. ψ^2). This was another new concept – the probability that a **certain quantum state exists**. No more exact answers, said Born. In atomic theory, all we get are *probabilities*.

Ground state of Hydrogen

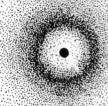

According to Bohr   According to Born

Two Kinds of Probability

On 10 August 1926, Born gave a paper at Oxford in which he clearly distinguished between the old and the new probabilities in physics. The old classical Maxwell-Boltzmann theory (see pages 20–25) had introduced microscopic co-ordinates in the kinetic theory of gases, only to eliminate them for **average** values based on probability due to ignorance. It had been impossible to calculate the exact values for so many particles.

BUT THE NEW THEORY GETS THE SAME RESULTS WITHOUT INTRODUCING AVERAGES AT ALL. THIS IS NOT *PROBABILITY* DUE TO IGNORANCE. THIS PROBABILITY IS ALL WE CAN EVER KNOW ABOUT AN ATOMIC SYSTEM.

Born had found a way to reconcile particles and waves by introducing the concept of probability. The wave ψ determines the likelihood that the electron will be in a particular position. Unlike the electromagnetic field, ψ has no physical reality.

Schrödinger's Cat . . . The Quantum Measurement Problem

About ten years after Born's papers, the notion of the probability superposition of quantum states was becoming generally accepted. Schrödinger, distressed that his own equation was being misused, created a "thought experiment" which he believed would demonstrate – once and for all – the absurdity of this concept.

Schrödinger imagined a bizarre experiment in which a live cat is placed in a box with a radioactive source, a Geiger counter, a hammer and a sealed glass flask containing deadly poison fumes. When a radioactive decay takes place, the counter triggers a device releasing the hammer which falls and breaks the flask. The fumes will then kill the cat.

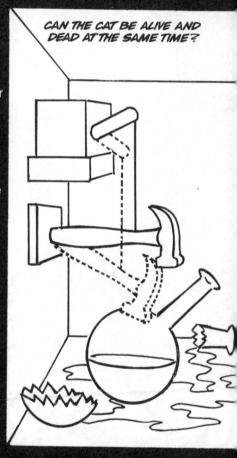

CAN THE CAT BE ALIVE AND DEAD AT THE SAME TIME?

SUPPOSE THE RADIOACTIVE SOURCE IS SUCH THAT QUANTUM THEORY PREDICTS A 50% PROBABILITY OF ONE DECAY PARTICLE EACH HOUR. AFTER AN HOUR HAS PASSED, THERE IS AN EQUAL PROBABILITY OF EITHER STATE . . . THE *LIVE* CAT STATE OR THE *DEAD* CAT STATE.

Quantum theory (with the Born interpretation) would predict that exactly one hour after the experiment began, the box would contain a cat that is **neither wholly alive nor wholly dead** but a mixture of the two states, the superposition of the two wave functions.

SEE, IT'S RIDICULOUS!

THE PROBABILITY INTERPRETATION OF MY WAVE FUNCTION IS NOT ACCEPTABLE!

Schrödinger thought he had made his point. Yet today, 60 years later, his so-called *paradox* is used to teach the concepts of quantum probability and the superposition of quantum states.

AS SOON AS WE LIFT UP THE LID OF THE BOX TO FIND OUT IF THE QUANTUM PREDICTION IS CORRECT, THE IMPASSE IS RESOLVED.

OUR ACT OF OBSERVATION COLLAPSES THE SUPERPOSITION OF THE TWO WAVE FUNCTIONS TO A SINGLE ONE, MAKING THE CAT DEFINITELY DEAD OR ALIVE.

Consciousness and the Collapsing Wave Function

The Hungarian-born physicist **Eugene Wigner** (1902–95), an expert on quantum theory and a Nobel Laureate, seems to have been one of the few bothered by **what actually causes the collapse of the wave function.**

THE CONSCIOUSNESS OF THE OBSERVER MAKES THE DIFFERENCE. WHEN WE BECOME CONSCIOUS OF SOMETHING, WE BRING ABOUT THE CRUCIAL COLLAPSE OF THE WAVE FUNCTION, SO THAT THE PERPLEXING MIXED STATES OF LIFE AND DEATH DISAPPEAR.

CRITICS ASKED IF AN AMOEBA COULD CAUSE A COLLAPSE, OR EVEN IF THE CAT'S OWN CONSCIOUSNESS COULD KEEP IT *REAL* THROUGHOUT THE EXPERIMENT. PHYSICS SEEMS UNRECOGNIZABLE FROM THE DAYS OF ISAAC NEWTON.

Wigner's has not been a popular explanation among physicists, nor even a serious concern. Quantum theory works. It gives practical answers to the most complicated theoretical questions. Those who use quantum theory as a work-a-day experience could not care less what causes the wave function to collapse!

Paul Adrian Maurice Dirac: Genius and Recluse

Having seen two alternative versions of the new quantum theory – the first by Heisenberg using matrix methods and the second dominated by Schrödinger's wave equation – now consider a third, developed independently by the English mathematician, Paul A.M. Dirac.

In the summer of 1925, Heisenberg gave a talk in the Kapitza Club at Cambridge, after which he gave his host Ralph Fowler a copy of the unpublished manuscript of his new pioneering theory. Fowler passed it on to his young graduate student Paul Dirac with the note, *See what you think of this*. Dirac took the instruction seriously.

Working alone – as he would do for the entire 44 years of his career in physics – Dirac saw that Heisenberg's was an important new departure.

IT IS CAPABLE OF RESOLVING THE DIFFICULTIES OF THE OLD QUANTUM THEORY OF *BOHR*, *EINSTEIN* AND *PLANCK*.

149

Dirac's Version of Quantum Mechanics

At first puzzled by the appearance of non-commuting quantities (where the product of two quantities depended on their order, so that **A x B** does not equal **B x A**), Dirac realised that **this** was the essence of the new approach. He quickly found a link to classical physics and used the new fundamental idea of non-commutability to develop his own version of *Quantum Mechanics*.

IN LESS THAN TWO MONTHS, I COMPLETED A THIRTY-PAGE PAPER WHICH I SENT TO HEISENBERG FOR HIS OPINION.

I HAVE READ YOUR EXTRAORDINARILY BEAUTIFUL PAPER ON QUANTUM MECHANICS WITH THE GREATEST INTEREST, AND THERE CAN BE NO DOUBT THAT ALL YOUR RESULTS ARE CORRECT. THE PAPER IS REALLY BETTER WRITTEN AND MORE CONCENTRATED THAN OUR EFFORTS HERE.

Heisenberg **and** Born were impressed! Dirac immediately became a member of *the club*, destined to be immortalized as one of the founders of quantum theory.

Dirac's Transformation Theory

But he was only just beginning. By November 1925, only four months after receiving the germ of the new mechanics, Dirac had written a series of four papers which attracted the attention of theoreticians everywhere, but particularly in Copenhagen, Göttingen and Munich, the main centres of quantum research. Putting these together as a thesis for the Cambridge faculty, they happily gave him a Ph.D.

Next, Bohr beckoned him to Copenhagen in September 1926. There Dirac completed another important paper on **transformation theory**.

I SHOWED THAT BOTH THE RECENTLY PUBLISHED WAVE MECHANICS OF ERWIN SCHRÖDINGER AND HEISENBERG'S ORIGINAL MATRIX MECHANICS COULD BE VIEWED AS SPECIAL CASES OF MY OWN MORE GENERAL FORMULATION. IN OTHER WORDS, *THEY ARE ALL EQUIVALENT.*

The Beginning of Quantum Electrodynamics

In Copenhagen and later in Göttingen, Dirac started working on problems of the emission and absorption of electromagnetic radiation, i.e. *light*. A quarter of a century earlier, Planck and Einstein had presented theoretical evidence that light consisted of particles, which today are called *photons*.

Despite overwhelming evidence during the 19th century for the wave model of light, Einstein had rekindled the controversy over the **duality of particles and waves**.

But common sense demanded that light must be one or the other. Dirac showed that quantum theory had the answer to this apparent paradox.

BY CONSISTENTLY APPLYING QUANTUM MECHANICS TO MAXWELL'S ELECTROMAGNETIC THEORY, I CONSTRUCTED THE FIRST KNOWN SPECIMEN OF A *QUANTUM FIELD THEORY*.

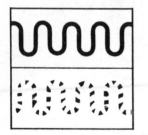

The concept of a continuous field, introduced by Faraday and others (remember those iron filings and the bar magnet in science class?), could now be broken up into bits in order to interact with matter, already known to consist of discrete entities like electrons, protons, etc. Dirac's new approach could treat light as waves **or** particles and give the right answers. Magic!

J.C. Polkinghorne (b. 1930), a former professor of theoretical physics at Cambridge who learned his quantum mechanics directly from Dirac, is today still impressed with this achievement, 70 years later. He offers a vivid metaphor . . .

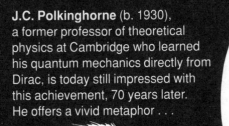

> DIRAC'S WAS A WELL-UNDERSTOOD FORMALISM WHICH IF INTERROGATED IN A PARTICLE-LIKE WAY GAVE PARTICLE BEHAVIOUR AND IF INTERROGATED IN A WAVE-LIKE WAY GAVE WAVE BEHAVIOUR.

> IT WAS AS IF SOMEONE HAD ASSERTED THAT IT WAS INCONCEIVABLE THAT A MAMMAL SHOULD LAY AN EGG AND THEN SUDDENLY A DUCK-BILLED PLATYPUS SHOWED UP.

Since this work of Dirac, the dual nature of light as wave and particle has been free of paradox for those who can follow the mathematics. After World War II, Dirac's pioneering work was carried forward by **Richard Feynman** (1918–88) and others.

> WE CALL OUR THEORY QUANTUM ELECTRODYNAMICS, OR **QED** FOR SHORT. IT DESCRIBES THE INTERACTION OF LIGHT AND MATTER WITH REMARKABLE ACCURACY.

The Dirac Equation and Electron Spin

International recognition did not change Dirac's habits greatly. Returning to Cambridge, he continued working intensely, almost always in the privacy of his room in the cloistered quadrangle of St. John's College. He was about to make another great discovery.

The wave mechanics of Schrödinger had taken centre stage and the ubiquitous *wave equation* dominated quantum theory (and still does for most practitioners). Schrödinger did not know about the electron's curious magnetic property, called *spin*. Consequently, he had not been able successfully to incorporate Einstein's relativity into his wave equation. Dirac did it for him in breathtaking style, using mainly aesthetic arguments.

PRESERVING THE SYMMETRY OF BOTH SPECIAL RELATIVITY AND QUANTUM MECHANICS, I GUESSED A NEW WAVE EQUATION FOR THE ELECTRON. IT SEEMS TO WORK.

The formula he found (now known as the *Dirac Equation*) not only gave the description of an electron moving close to the speed of light, but predicted **without any *ad hoc* hypotheses** that the electron had a spin of one-half, as was known from experiments.

The Prediction of Anti-Matter

Remarkably, Dirac's equation also dictated the existence of a **positively charged electron**, the opposite charge to every electron which had previously been observed.

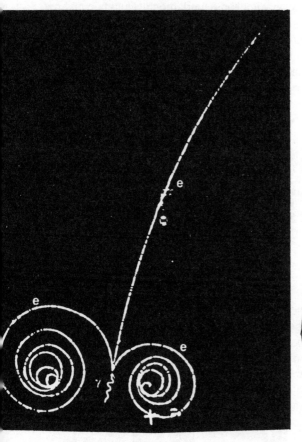

THIS WAS THE FIRST CLUE THAT THERE MIGHT BE SUCH A THING AS ANTI-MATTER, PARTICLES WITH MASS AND SPIN IDENTICAL TO ORDINARY MATTER, BUT WITH OPPOSITE ELECTRIC CHARGE.

This prediction was verified a few years later, when anti-electrons, now called *positrons*, were discovered by **Carl Anderson** in a cloud chamber at Caltech in 1932. Dirac had opened up a broad area of anti-particle physics.

Only a year after the positron was observed, Dirac received the Nobel Prize for 1933, awarded jointly to him and Schrödinger for their work on quantum theory. Let's go back to 1926–7 . . .

The Uncertainty Principle

In 1927, Heisenberg made a second major discovery, one as important as his discovery of matrix mechanics. Driven by his positivist belief that only measurable quantities should be part of any theory, Heisenberg realized that quantum theory implied a fundamental limitation on how accurately certain pairs of physical variables could be measured simultaneously. Here's what he did.

Recall the non-commutability of the two variables – **position** and **momentum (pq – qp = h/2πi)** . . .

I SHOWED THAT THERE IS NO WAY OF ACCURATELY PINPOINTING THE *EXACT POSITION* OF A SUB-ATOMIC PARTICLE, UNLESS YOU ARE WILLING TO BE QUITE UNCERTAIN ABOUT THE PARTICLE'S MOMENTUM.

ALSO, THERE IS NO WAY TO PINPOINT THE PARTICLE'S *EXACT MOMENTUM* UNLESS YOU ARE WILLING TO BE QUITE UNCERTAIN ABOUT ITS POSITION. TO MEASURE BOTH ACCURATELY AT THE SAME TIME IS IMPOSSIBLE.

A quantitative relationship for this uncertainty was easily derived by **estimating the imprecision** in a simultaneous measurement of position and momentum. To locate or "see" precisely any object, the illuminating radiation must be **significantly smaller than the object itself**. For an atomic electron, this means waves much smaller than the ultraviolet, as the diameter of the entire hydrogen atom is only a fraction of the wavelength of visible light.

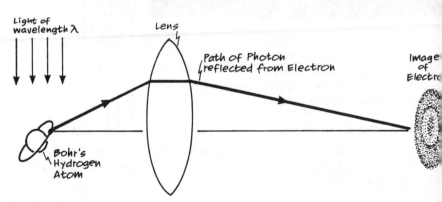

Light of wavelength λ

Lens

Path of Photon reflected from Electron

Image of Electron

Bohr's Hydrogen Atom

Heisenberg's Gamma-Ray Microscope

To study the problem, Heisenberg chose a hypothetical microscope using **gamma rays**, which are very short but carry considerable momentum. Thus, the path of the electron is not smooth and continuous, but herky-jerky due to bombardment by the gamma-ray photons. George Gamow's famous drawing of Heisenberg's hypothetical set-up is shown on this page. Bohr helped Heisenberg clarify this part of the derivation.

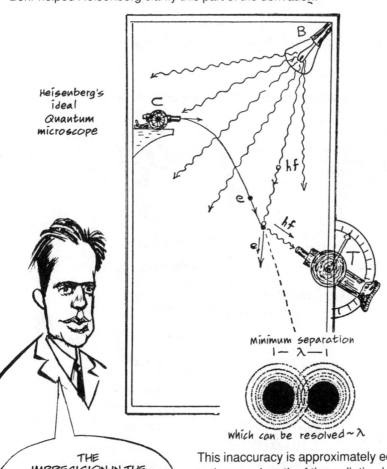

Heisenberg's ideal Quantum microscope

B

C

hf

e

hf

e'

T

Minimum separation

$|— \lambda —|$

which can be resolved $\sim \lambda$

THE IMPRECISION IN THE POSITION MEASUREMENT OF AN OBJECT UNDER HIGH OPTICAL MAGNIFICATION, E.G. A MICROSCOPE, IS LIMITED BY DIFFRACTION, WHEN INTERFERENCE PATTERNS OVERLAP.

This inaccuracy is approximately equal to the wavelength of the radiation being used, as shown in the sketch. Thus, the **imprecision in the position measurement** is $\Delta \chi \sim \lambda$. (N.B. χ is being used for position instead of **q**, and $\sim$ means "approximately equal to".) 157

Correspondingly, the minimum **imprecision in the momentum measurement** is approximately equal to the momentum imparted to the electron by a **single photon** used to illuminate the particle, the smallest disturbance possible. From the de Broglie/Einstein relation, **Δp ~ h/λ**, Heisenberg obtained the imprecision in the momentum. Multiplying the two inaccuracies together, Heisenberg showed that the product, **Δx Δp** will always be greater than or equal to (≥) a certain amount . . .

de Broglie relation

$$(\Delta x)(\Delta p) \geq (\lambda)(h/\lambda) \geq h \quad OR \dots \quad \Delta x \, \Delta p \geq h$$

from diffraction limit

This is **Heisenberg's Uncertainty Principle** (HUP), which states that . . .

In der vorliegenden Arbeit werden zunächst exakte Definitionen Geschwindigkeit, Energie usw. (z. B. des Elektrons) aufgestellt, Quantenmechanik Gültigkeit bei und es wird gezeigt, daß jugierte Größen simultan nur mi werden können (§ 1). Diese I Auftreten statistischer Zusamm matische Formulierung gelingt n den so gewonnenen Grundsätze Vorgänge aus der Quantenmec Erläuterung der Theorie werden

Though we don't notice HUP in our everyday experience with the gross macroscopic world, the wave/particle duality defeats the atomic experimentalist who seeks perfection. But many believe there are serious **philosophical consequences** for us all in this idea.

The Breakdown of Determinism

In the late 18th century, the French philosopher **Pierre Simon de Laplace** (1749–1827) stated the *Principle of Determinism:*

. . . IF AT ONE TIME, WE KNEW THE POSITIONS AND MOTION OF ALL THE PARTICLES IN THE UNIVERSE, THEN WE COULD CALCULATE THEIR BEHAVIOUR AT ANY OTHER TIME, IN THE PAST OR FUTURE.

(TRANSLATED FROM THE FRENCH BY STEPHEN HAWKING)

HUP DESTROYS THE FIRST PREMISE OF THIS STATEMENT, IN THAT WE CANNOT KNOW THE PRECISE POSITION *AND* MOTION OF A PARTICLE *AT ANY TIME*. THUS, DETERMINISM CANNOT BE ACCEPTED CONGRUENTLY WITH *HUP*.

This conclusion has its critics who say that such a relation, based on the atomic world, cannot legitimately be raised to a universal law. This was answered eloquently some years ago by **Victor Weisskopf** (b. 1908), a Hungarian physicist who attended many meetings in the 1930s at Bohr's institute.

THE UNCERTAINTY PRINCIPLE HAS MADE OUR UNDERSTANDING OF NATURE RICHER, NOT POORER. IT LIMITS THE APPLICABILITY OF CLASSICAL PHYSICS TO ATOMIC EVENTS TO MAKE ROOM FOR NEW PHENOMENA LIKE THE WAVE/PARTICLE DUALITY. TO QUOTE FROM *HAMLET:*

"*THERE ARE MORE THINGS IN HEAVEN AND EARTH, HORATIO, THAN ARE DREAMT OF IN YOUR PHILOSOPHY.*"

But no one could have dreamt what was in the other "Great Dane" **Bohr**'s philosophy in the spring of 1927.

159

Complementarity

On a skiing holiday in Norway in 1927, Bohr found what he believed to be the central core of understanding quantum mechanics, the wave/particle duality. But he had a new point of view.

ALTHOUGH WAVE AND PARTICLE BEHAVIOUR OF AN OBJECT MUTUALLY EXCLUDE EACH OTHER, BOTH ARE NECESSARY FOR THE FULL UNDERSTANDING OF THE OBJECT'S PROPERTIES. I CALL THIS NEW SITUATION, *COMPLEMENTARITY*.

But as a quantum physicist, I would say . . .

A classical physicist would say . . .

IF TWO DESCRIPTIONS ARE MUTUALLY EXCLUSIVE, THEN AT LEAST ONE OF THEM MUST BE WRONG.

WHETHER AN OBJECT BEHAVES AS A PARTICLE OR AS A WAVE DEPENDS ON YOUR CHOICE OF APPARATUS FOR LOOKING AT IT.

Particle Detectors

Source

observing photon with a particle detector

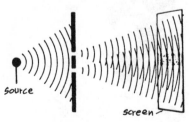

Source

screen

observing Photon with a Wave detector

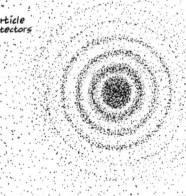

Electron reveals Wave and Particle properties

The Copenhagen Interpretation

After arguing with Heisenberg for weeks over this concept, Bohr began to bring together the various parts of quantum theory into a consistent whole. He combined various aspects of Heisenberg's work – matrix mechanics and the uncertainty principle – with Born's probability interpretation of the Schrödinger wave equation and his own **complementarity**.

This was another new concept, focusing on the quantum measurement problem and its all-important connection to classical physics. This collection of ideas became known as the **Copenhagen Interpretation** (CHI).

Como, Italy, September 1927

After struggling for months to articulate his thinking on all aspects of quantum theory, Bohr presented a lecture at Como to most of Europe's best physicists in September 1927. Free from Einstein's critical eye and ear (he would not set foot in fascist Italy), Bohr described in detail the **Principle of Complementarity** for the first time.

SUPPOSE ONE SET OF EXPERIMENTAL EVIDENCE CAN ONLY BE INTERPRETED ON THE BASIS OF *WAVE* PROPERTIES AND ANOTHER SET ONLY ON THE BASIS OF *PARTICLE* PROPERTIES. THESE TWO SETS OF EVIDENCE ARE NOT CONTRADICTORY.

SINCE THE EVIDENCE WAS OBTAINED UNDER DIFFERENT EXPERIMENTAL CONDITIONS, IT CANNOT BE COMBINED IN A SINGLE PICTURE BUT MUST BE REGARDED AS *COMPLEMENTARY*.

The Solvay Conference, October 1927

At the end of October 1927, only weeks after the Como meeting, Bohr arrived at the Metropole Hotel in Brussels for the historic Solvay Conference highlighted at the beginning of this book.

THIS TIME, EINSTEIN WILL BE PRESENT AND I AM EAGER TO HEAR WHAT HE WILL HAVE TO SAY.

Einstein wanted a theory to describe the **thing itself** and not the **probability** of its occurrence. Yet Bohr was confident that Einstein would accept his interpretation, which was tied to experiments. This was the method Einstein himself had used in defending his theory of special relativity, which also challenged common sense.

But to Bohr's shock and disappointment, Einstein announced . . .

I DO NOT LIKE THE PROBABILITY THEORY AND BELIEVE THE PATH FOLLOWED BY BORN, HEISENBERG AND YOURSELF IS ONLY TEMPORARY, OF HEURISTIC VALUE, SO TO SPEAK.

Einstein set out to demolish CHI by attacking the "distasteful" uncertainty principle on which it was based. He used ingenious **thought experiments**, trying to contradict Heisenberg's law. But each time Bohr found a flaw in Einstein's scheme and refuted the argument.

163

Einstein's Box of Light

Three years later, at the next Solvay Meeting, the most serious challenge occurred. Einstein believed that he had finally found a case where HUP was violated. He described a box full of light and suggested that both the **energy** of a single photon and the **time** it was emitted could be determined precisely. Time and energy were, in principle, another pair of variables governed by HUP.

FIRST, THE BOX CAN BE WEIGHED, AFTER WHICH A SINGLE PHOTON CAN BE RELEASED AT A PARTICULAR INSTANT THROUGH A SHUTTER OPERATED BY CLOCKWORK INSIDE THE BOX.

THEN THE BOX CAN BE WEIGHED AGAIN. KNOWING THE CHANGE IN MASS, THE ENERGY OF THE PHOTON CAN BE CALCULATED FROM MY EQUATION, $E = MC^2$.

THE ENERGY CHANGE WOULD THEN BE KNOWN, AS WOULD THE PRECISE TIME WHEN THE PHOTON WAS EMITTED. SO, THAT'S THE END OF YOUR UNCERTAINTY PRINCIPLE!

A Sleepless Night

Was Bohr stumped? Apparently he lay awake all night trying to work out what was wrong with the experiment before the answer finally appeared. Next morning, he produced a drawing of the box of light. Then, much to Einstein's chagrin, Bohr refuted his "light box" argument.

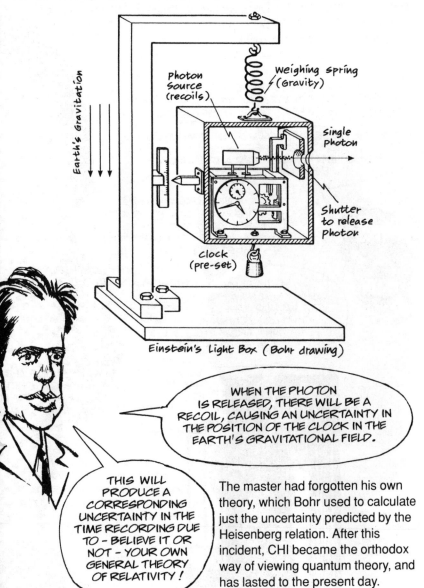

Einstein's Light Box (Bohr drawing)

The master had forgotten his own theory, which Bohr used to calculate just the uncertainty predicted by the Heisenberg relation. After this incident, CHI became the orthodox way of viewing quantum theory, and has lasted to the present day.

The EPR Paradox

But did Einstein give up? Not exactly. Five years later, after Hitler's rise to power had dispersed European physicists all over the world, Einstein ended up at the Institute for Advanced Study in Princeton, New Jersey. With two younger colleagues, **Boris Podolsky** (1896–1966) and **Nathan Rosen** (b. 1909), he developed another challenge to Bohr that was **not based on the uncertainty principle**. It is known as the **EPR paradox** after its authors' names.

IT IS POSSIBLE TO OBTAIN A PAIR OF PARTICLES, SAY ELECTRONS, IN A SO-CALLED *SINGLET STATE* WHERE THEIR SPINS CANCEL EACH OTHER TO GIVE A TOTAL SPIN OF ZERO. LET US SUPPOSE THESE PARTICLES *A* AND *B* MOVE WIDELY APART, AFTER WHICH THE SPIN OF *A* ALONG ONE DIRECTION IS MEASURED AND FOUND TO BE IN THE "UP" STATE.

BECAUSE THE TWO SPINS MUST CANCEL TO ZERO, IT FOLLOWS THAT PARTICLE *B* ALONG THE SAME DIRECTION MUST HAVE SPIN "DOWN".

IN CLASSICAL PHYSICS, THIS WOULD NOT BE A PROBLEM AT ALL. ONE WOULD JUST CONCLUDE THAT PARTICLE *B* ALWAYS HAD SPIN "DOWN", FROM THE TIME OF THE SEPARATION.

Podolsky

Einstein leaves Germany for good, 1933.

The Locality Principle

HOWEVER, ACCORDING TO *CHI*, THE SPIN OF *A* HAS NO DEFINITE VALUE UNTIL IT IS MEASURED, AT WHICH POINT IT MUST PRODUCE AN INSTANTANEOUS EFFECT AT *B*, COLLAPSING ITS SPIN WAVE FUNCTION INTO THE OPPOSITE OR "DOWN" STATE.

THIS BIZARRE SITUATION DEMANDS *ACTION-AT-A-DISTANCE* OR *FASTER THAN LIGHT* COMMUNICATION, NEITHER OF WHICH IS ACCEPTABLE.

Rosen

Einstein and his colleagues were convinced they had demonstrated the existence of hidden variables (*elements of reality*) which quantum theory fails to take into account, thus showing the theory to be **incomplete**.

The big issue here was that of Einstein's **separateness**, i.e. his locality principle . . .

IF TWO SYSTEMS ARE IN ISOLATION FROM EACH OTHER FOR SOME TIME, THEN A MEASUREMENT ON THE FIRST CAN PRODUCE *NO REAL CHANGE* ON THE SECOND.

DON'T FORGET MY SPECIAL RELATIVITY—NOTHING TRAVELS FASTER THAN LIGHT!

Bohr and Non-Locality

This separateness or locality was not allowed, said Bohr. He immediately reminded Einstein (and the world) what CHI had always asserted . . .

*Quantum mechanics does not permit a separation between the observer and the observed. The two electrons **and** the observer are part of a single system. The EPR experiment does not demonstrate the incompleteness of quantum theory, but the naiveté of assuming **local** conditions in atomic systems. Once they have been connected, atomic systems never separate.*

The big question was whether this remarkable property of non-locality could ever be experimentally tested. Or could the existence of Einstein's **separateness** be proven instead?

Bell's Inequality Theorem

For thirty years after EPR, very little progress was made on this important question, until a Belfast physicist, **John S. Bell** (1928–90) took a one-year leave from CERN (European Centre for Nuclear Research). He developed an ingenious **inequality principle** to test the questions raised by the paradox.

THE TEST IS BASED ON CORRELATED *PHOTONS* (INSTEAD OF ELECTRONS) IN WHICH THE POLARIZATION OF THE LIGHT IS DETECTED INSTEAD OF SPIN. BUT THE PRINCIPLES ARE THE SAME: *HOW DO CHANGES IN A AFFECT B?*

To derive his inequality, Bell used certain facts and ideas with which everyone could agree, except for . . . Einstein's condition of locality, which he assumed to be true.

Now, if experiments showed that the **inequality was violated**, this would mean that one of the premises in his derivation was false. Bell chose to interpret this to mean that nature is non-local.

Experiments by **John Clauser** and others at Berkeley in 1978 and, in particular, by **Alain Aspect**'s group in Paris in 1982, indicated experimental verification of the violation of Bell's inequality.

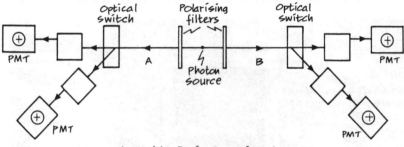

PMT = Photomultiplier Tube

Aspect's Paris Experiment, 1982

This means that in spite of the local appearances of phenomena, our world is actually supported by an invisible reality which is unmediated and **allows communication faster than light, even instantaneously.**

Interactions under Non-Local Reality
1. The interaction does not diminish with distance.
2. It can act instantaneously (faster than the speed of light).
3. It links up locations without crossing space.

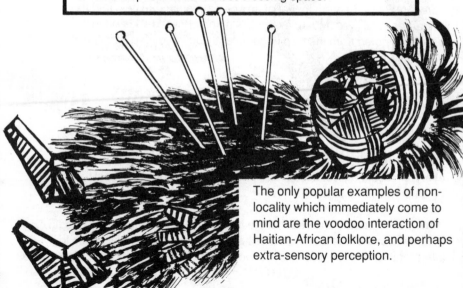

The only popular examples of non-locality which immediately come to mind are the voodoo interaction of Haitian-African folklore, and perhaps extra-sensory perception.

An Undiscovered World

This would seem to be a most remarkable aspect of nature, and a discovery resulting from the application of quantum theory. Bell's work, which should apply to any fundamental theory of nature (i.e. not just quantum theory), could turn out to be one of the most important theoretical ideas of this century.

In spite of much enthusiasm in the last decade, there now appear to be certain loopholes in experiments like Aspect's, based on the statistical analysis of hundreds of measurements. These loopholes have reverted the proof of Bell's theorem to that of an open question. Einstein and the EPR paradox still lives! Much research is going on world-wide on this question, as noted from the web page recently down-loaded from the Internet.

On
Tuesday 21 November 1995, 16.00 h precisely **at the UvA** ~~ ·
Thursday 7 December 1995, 16.00 h precisely **at** N¹⁴··

Philippe Eberhard reviewed the ⸺
at the Univeristy of Am⸺·
were not met ⸺
At N¹⁴··

the EPR Paradox and Bell's Inequality

rinciple

1935 Albert Einstein and two colleagues, Boris Podolsky and Nathan Rosen (EPR) developed a thought xperiment to demonstrate what they felt was a lack of completeness in quantum mechanics. This so-called EPR paradox" has led to much subsequent, and still on-going, research. T⸺ article is an introduction to EPR, Bell's inequality, and the real experiments which have attempted to ⸺ ⸺sting issues raised by this discussion.

One of the principal features of quantum mechanics is that not all the cla⸺

World Wide Web Worm: Future
⸺er 1995, 16.00 h precisely

NI☒EF

WWWW - the WORLD WIDE WEB WORM
File Edit View Go Bookmarks Options Directory

Location: http://www.cs.colorado.edu/home/mcbryan/WWWW.html

WWWW - the WORLD WIDE WEB WORM

Best of the Web '94 - Best Navigational Aid, Oliver McBryan
Last Run Sept 5 Users: 2,000,000 per month.
Introduction, Definitions, Search Examples, Failure, Register a Resource, WWWW Pages

1. Search only in Titles of citing documents
2. Search only in Names of citing documents
3. Search all Citation Hypertext
4. Search all Names of Cited URL's

Quantum Theory and the New Millennium

The famous exchange depicted in the photo on this page does not represent Einstein's **most serious** challenge to Bohr's interpretation of quantum theory. Schrödinger's waves and Heisenberg's uncertainty principle do work! But the EPR paradox is another matter.

It is true that the experiments on correlated photons in 1982 (Aspect, et al.) seemed to confirm violations of Bell's Theorem, suggesting that nature is non-local. The matter appeared to be settled.

But can **non-locality** really be true? Can we live with the preposterous concept of **action-at-a-distance** (voodoo, ESP, etc)?

Today not everyone agrees that the correlation experiments are conclusive. So, where does that leave us now?

John Archibald Wheeler, Quantum Physicist

The one man living today to answer this question is **John Wheeler** (b. 1911), Emeritus Professor of Physics at Princeton University. Wheeler has been at the cutting edge of 20th century physics – relativistic cosmology and quantum theory – for over 60 years. He is well-known for his endless efforts to comprehend all aspects of the quantum formalism. His work has emphasized the central role of the observer in *creating reality*.

SOME OF US JUST CAN'T ACCEPT ALL THE EXPLANATIONS IMPLICIT IN *CPI*, ESPECIALLY NON-LOCALITY. COULD IT BE THAT EINSTEIN IS RIGHT ...AGAIN?

The author visited John Wheeler at Princeton on a snowy day in December 1995

FOR THE *EPR* PARADOX, REMEMBER: WE HAVE NO RIGHT TO ASK WHAT THE PHOTONS ARE DOING DURING THEIR TRAVEL. *NO ELEMENTARY PARTICLE IS A PHENOMENON UNTIL IT IS REGISTERED.* I MUST SAY IN EVERYDAY CONTEXT, QUANTUM THEORY IS UNSHAKEABLE, UNCHALLENGEABLE, UNDEFEATABLE – IT'S *BATTLE TESTED.*

A Final Word

Wheeler wrote to the author recently . . .

December, 2000, is the 100th anniversary of the greatest discovery ever made in the world of physics, the quantum. To celebrate, I would propose the title, "The Quantum: The Glory and the Shame". Why glory? Because there is not a branch of physics which the quantum does not illuminate. The shame, because we still do not know "how come the quantum?".

173

FURTHER READING

Quantum theory cannot be *explained*. Physicists and mathematicians from Niels Bohr to Roger Penrose have admitted that it doesn't make sense. What one *can* do is discover how the ideas developed and how the theory is applied. Our book has concentrated on the former. Other recommendations are listed below.

Development of Quantum Theory

The Quantum World, J.C. Polkinghorne. Penguin 1990. Excellent though condensed read by a man who learned the subject from Dirac.

Thirty Years That Shook Physics, George Gamow. Doubleday 1966. A gem by the humorous physicist/cartoonist who first applied the discoveries of quantum theory in the 1930s. Available as a Dover paperback.

In Search of Schrödinger's Cat, John Gribbin. Bantam 1984. Until now, this was the best lay person's guide to how the theory emerged. Gives examples of theory's application, describes Feynman's quantum electrodynamics and summarizes modern interpretation at that time.

Taking the Quantum Leap, Fred Alan Wolf. Harper and Row 1989. Lively presentation of the basic yet surprising ideas of quantum theory.

Life and Work of Principal Players

The Dilemmas of an Upright Man, J.L. Heilbron. Univ. of California Press 1986. Sympathetic and thorough biography of Max Planck, who discovered the quantum.

Subtle is the Lord, Abraham Pais. Oxford University Press 1982. Of the dozens of biographical writings on Einstein, this is the definitive treatment.

Niels Bohr's Times, Abraham Pais. Oxford University Press 1991. Unusual storytelling reveals a man who groped his way through most of 20th century atomic physics as its leading authority.

Physics and Philosophy, Werner Heisenberg. Harper 1958. Discussion of the Copenhagen Interpretation and its importance to philosophy by the discoverer of matrix mechanics and the uncertainty principle with thirty years' perspective.

The Restless Universe, Max Born. Dover 1951. Easy to read classic on 20th century physics, including explanation of the statistical aspects of quantum theory. Corner flip pages demonstrate time sequences.

Matter and Light, Louis de Broglie. Norton 1939 (also Dover paperback). The French prince's point of view as he remembers it.

Schrödinger: Life and Thought, Walter Moore. Cambridge University Press 1989. Acclaimed biography of the Austrian polymath, warts and all.

Beyond the Atom: Philosophical Thoughts of Wolfgang Pauli, K.V. Laurikainen. Springer-Verlag 1985. Thoughts of the cynical man who dreamt up the exclusion principle and who once described a theory as being so bad . . . *it wasn't even wrong!*

Directions in Physics, Paul Dirac. Wiley 1978. A set of lectures given by Dirac which includes his view of the unfinished work of fundamental theorists.